International Scientific Conference Arctic: History and Modernity 2018

IOP Conference Series: Earth and Environmental Science
Volume 180

St. Petersburg, Russia
18 – 19 April 2018

ISBN: 978-1-5108-6863-2
ISSN: 1755-1307

Printed from e-media with permission by:

Curran Associates, Inc.
57 Morehouse Lane
Red Hook, NY 12571

Some format issues inherent in the e-media version may also appear in this print version.

Printed by Curran Associates, Inc. (2018)

For permission requests, please contact the Institute of Physics
at the address below.

Institute of Physics
Dirac House, Temple Back
Bristol BS1 6BE UK

Phone: 44 1 17 929 7481
Fax: 44 1 17 920 0979

techtracking@iop.org

Additional copies of this publication are available from:

Curran Associates, Inc.
57 Morehouse Lane
Red Hook, NY 12571 USA
Phone: 845-758-0400
Fax: 845-758-2633
Email: curran@proceedings.com
Web: www.proceedings.com

Table of contents

Volume 180

Arctic: History and Modernity

18–19 April 2018, Saint Petersburg, Russian Federation

Accepted papers received: 25 July 2018
Published online: 10 August 2018

Preface

Papers

Historical geography of the Arctic region

Ecology and protection of the Arctic environment

Social geography of the Arctic

Energy and clean technologies for the Arctic

Economy of the Arctic environment

IOP Conf. Series: Earth and Environmental Science **180** (2018) 011001 doi:10.1088/1755-1315/180/1/011001

3rd International Scientific Conference

ARCTIC: HISTORY AND MODERNITY

18-19 April 2018

St. Petersburg
2018

IOP Conf. Series: Earth and Environmental Science **180** (2018) 011001 doi:10.1088/1755-1315/180/1/011001

PREFACE

The theme of the Arctic, where geopolitical and economic interests of many countries intersect now, is truly relevant. In the Arctic ocean, Canada, Denmark, Norway, the United States and the Russian Federation (USSR) have a 370-kilometer exclusive economic zone off their shores. A part of the Arctic was owned by these countries in the 20s of the last century. In addition, more than 20 countries declare their readiness to develop deposits on the Arctic shelf, among them - India, China, South Korea, Brazil, Germany, Japan.

Against the background of increased interest in the Arctic in the modern world, it is important to report on a wide range of issues to a wide audience. These issues characterize historical, business, socio-political and other activities of Russia in the Arctic region. The issues of interaction between the Arctic countries in the conditions of modern realities and public diplomacy are also important.

The annual international scientific conference "The Arctic: history and modernity" is designed to unite the efforts of technical and humanitarian specialties representatives to develop and apply an interdisciplinary approach to the Arctic issue.

The third international scientific conference "The Arctic: history and modernity" was held on April 18-19, 2018 at Peter the Great St. Petersburg Polytechnic University. The co-organizers of the conference were the Institute of History of Natural Science and Technology named after S. I. Vavilov and the Institute of Robotics and Technical Cybernetics.

For the first time, co-organizers of the conference were: Darmstadt Technical University, Institute of Europe, RAS. In addition, some new partners were involved: the Museum of the World Ocean " Krasin Icebreaker", the Museum of the Arctic and Antarctica, the Project office of the Arctic development, Informational and analytical magazine "Regional energy science and energy supply", Murmansk Regional Museum, as well as the Murmansk State Technical University (MSTU).

The conference heard a wide range of views expressed and advocated by representatives of different countries. The conference was attended by 257 participants from Russia, France, Norway, Finland and India. 147 reports were heard. The "Geography" of the participants included such cities as St. Petersburg, Moscow, Murmansk, Apatity, Petrozavodsk, Arkhangelsk, Salekhard, Yakutsk and Astrakhan, Paris, Oslo, Helsinki, Boston, New Delhi.

The directions of the conference work were as follows:

- History of research and development in the Arctic
- Military security and strategic stability in the Arctic
- Problems of environmental and technical safety in the Arctic region
- International cooperation in the Arctic
- Problems and prospects of economic development of the Arctic zone of the Russian Federation
- Exploration and mining in the Arctic
- Historical and cultural heritage of the Arctic
- Tourism and its developmental prospects in the Arctic region
- Social geography of the Arctic
- Energy and clean technologies for the Arctic
- Arctic Robotics
-

The most interesting papers were selected for publication in this collection:

• Social geography of the Arctic
* Historical geography of the Arctic region
* Ecology and protection of the Arctic environment
* Energy and clean technologies for the Arctic
• Economy of the Arctic environment

Program Committee

Chairman

Y S VASILYEV- scientific supervisor of Peter the Great St. Petersburg Polytechnic University, Russia

Members

- N A ASHCHEULOVA — Director of St. Petersburg Branch of the Institute of History of Science and Technology of the Russian Academy of Sciences (co-chair), Russia
- N I ALMAZOVA – Director of Humanities Institute of Peter the Great St. Petersburg Polytechnic University (vice-chairperson), Russia
- N I DIDENKO – Professor, International graduate school of management, of Peter the Great St. Petersburg Polytechnic University, head of the «System dynamics» laboratory (vice-chairman), Russia
- S G LAGUSHKIN – Assistant to the Chairman of the Board of Trustees of Peter the Great St. Petersburg Polytechnic University (Executive Secretary of the Program Committee), Russia
- WANG QI – Executive Director of the Institute of strategic cooperation between China and Russia of Tsinghua University, China
- WANG TSI – Executive Director of the Institute of strategic cooperation between China and Russia, Tsinghua University, China
- V A GAISENOK – Rector of Republican Higher Institute of the Belarusian State University, Belarus
- V G GUSAKOV — Chairman of the Presidium of the National Academy of Sciences of Belarus, the Republic of Belarus (on agreement)
- V P ZHURAVEL — Senior researcher, Centre for Northern Europe, Institute of Europe, RAS, Russia
- V V IVANTER — Director, Institute of Economic Forecasting RAS, Scientific Supervisor of the laboratory «System Dynamics» of Peter the Great St. Petersburg Polytechnic University Russia
- A N KASHEVAROV — Professor of the Graduate School of Social Sciences of Peter the Great St. Petersburg Polytechnic University, Russia

- SV KULIK — the Director of the Graduate School of Social Sciences Peter the Great St. Petersburg Polytechnic University, Russia
- A V LOPOTA — Director, the chief designer of State Research Center of Russian Federation «Central research and experimental design institute of robotics and technical cybernetics», Russia
- NI VEJDOU — Chairman of the Academic Council of the Institute of strategic cooperation between China and Russia, Tsinghua University, China
- ALFRED NORDMANN — Professor of the Technical University of Darmstadt, Germany
- B N PORFIRIEV – Deputy Director, Institute of Economic Forecasting RAS, senior researcher of the «System Dynamics» laboratory at Peter the Great St. Petersburg Polytechnic University, Russia
- MASAMISI SASAKI — Professor at the University of Chuo University, Japan
- D F SKRYPNYUK — Professor, doctor of science in the International Graduate School of management of Peter the Great St. Petersburg Polytechnic University
- M B TENDLER – Professor of Alvenovskaya laboratory for thermonuclear studies of Royal Institute of Technology in Stockholm, Sweden
- E V TOROPOVA — Director of the Institute for Humanities NovSU in the name of Yaroslav The Wise, Russia
- A L TOLSTIK — BSU Vice Rector, Republic of Belarus
- LASLO UNGVARI – Rector of the University of Applied Sciences Wildau, Germany
- MICHAEL FRIEDMAN – Professor at the University of Regensburg, Germany
- K V CHISTYAKOV — Director of the Institute of Earth Sciences, Vice President of the Russian Geographical Society, Russia
- CHEN KAJKE — Researcher of the Institute of modern history under the Chinese Academy of Social Sciences of Tsinghua University, China
- H SHAMBURG – Professor of the Technical University of Hamburg, Germany
- JÜRGEN JAGER — Head of the department «World economy and international finance» at the University of Regensburg, Germany

Organizing Committee

Chairman

V V SERGEEV — Rector of Science of Peter the Great St. Petersburg Polytechnic University, Russia

Members

- S V KULIK — Director of the Graduate School of Social Sciences, Peter the Great St. Petersburg Polytechnic University, Russia (Vice-Chairman)
- E A SAMYLOVSKAYA — Assistant Professor of the Graduate School of Social Sciences, Peter the Great St. Petersburg Polytechnic University, Russia
- B N KOVALYOV — Professor of the Graduate School of Social Sciences, Peter the Great St. Petersburg Polytechnic University, Russia
- V V KORABLYOV — Adviser in the Administration of Peter the Great St. Petersburg Polytechnic University, Russia
- A A MIKHAILOV — a professor in the Graduate School of Social Sciences, Peter the Great St. Petersburg Polytechnic University, Russia
- A Y SKRYDLOV — researcher of St. Petersburg branch of the Institute for the History of Science and Technology in the name of S.I. Vavilov, RAS, Russian
- A B TESLYA — Assistant Professor of the International Graduate School of Management, Peter the Great St. Petersburg Polytechnic University, Russia

IOP Conf. Series: Earth and Environmental Science **180** (2018) 011002 doi:10.1088/1755-1315/180/1/011002

Peer review statement

All papers published in this volume of *IOP Conference Series: Earth and Environmental Science* have been peer reviewed through processes administered by the proceedings Editors. Reviews were conducted by expert referees to the professional and scientific standards expected of a proceedings journal published by IOP Publishing.

The Russian Geographical Society and the Polar Studies in the Second Half of the 19th Century

N G Sukhova and A Yu Skrydlov

Saint Petersburg Branch of the Institute of the History of Science and Technologies of the Russian Academy of Sciences, Saint Petersburg, Russia

askrydlov@gmail.com

Abstract. Acknowledgement of the necessity to study the northern regions of the Russian Empire in government circles was preceded by a long period of individual enthusiasts and industrialists' activities. They knew about the natural resources of the North. Local industrialists had understood almost before scientists polar regions economic importance. They tried to convey the idea of the necessity for scientific research in the North to the general public. The authors mainly relying on the published sources analyze the Russian Geographical Society (RGS) role in the polar research history. The article examines RGS members' attitude to the idea of organizing polar expeditions, its contacts with foreign polar travellers, determines the main results of the expeditions organized by RGS. Those studies led the authors to a conclusion that in the second half of the 19th century, RGS to some extent obtained the role of the platform hosting a discussion prospects for polar research. The Society has rarely dealt with the organization of polar research itself. Articles on problems connected to the observations results in the polar regions appeared on its periodicals pages. Travellers' reports were heard at RGS meetings. Various achievements in North studies were discussed. At the same time, RGS participation in the work of the First International Polar Year (1882-1883), the publication of a large number of papers on the polar travels history, the discussion of A. Nordenskiold and F. Nansen's expeditions results helped to draw attention to the issue of exploring the Arctic not only by scientists but also by state officials

1. Introduction

Perception of the natural conditions to the Arctic Ocean and its surroundings throughout the 19th century were quite primitive. Back to the 1860-s, it was unknown whether the whole ocean was covered with solid ice or the pole might be accessed by ships. The German cartographer A. Peterman's hypothesis about Greenland peninsula stretching to the Bering Strait prevailed until the end of the century. Simultaneously the Russian society didn't posses any common opinion regarding demand for polar research. The Russians didn't realize importance for it. Mainly entrepreneurs considered such studies necessary and sought to draw attention to the problem.

2. Main Body

In 1845, RGS was created and the first expedition, organized by it was sent to the Northern Urals. Although by the beginning of the 19th century the Ural Mountains were the most famous mountain range in Russia, it was still unclear where exactly the "Ural Ridge" ends in the North and in the South.

The Ural Expedition was organized by RGS following Academician, Director of the Pulkovo Observatory V.Y. Struve's initiative. The expedition was headed by St. Petersburg University Professor E.K. Hoffman, a participant of the expedition was an astronomer, Kazan University Professor M.A. Kovalevsky.

Summer 1847, 1848 and 1850 astronomical, geological, botanical and zoological observations, geodetic surveys and heights measurements were conducted in the Urals. During the expedition, the Ural Mountains highest peak named Konstantinov Stone (in honor of the RGS Chairman) was discovered. The direction of the Northern Urals and the ratio between the North Urals and the Pai-Khoi Ridge were revealed. The Ural expedition results were published in two volumes "The Northern Urals and the Pai-Khoy Coast Range" (St. Petersburg, 1854, 1856). They were highly appreciated by contemporaries. The expedition members were awarded by the Academy of Sciences and RGS.

Unfortunately, for a long time afterwards RGS didn't organize similar expeditions to the polar regions. The study of the North was done only by the enthusiasts. One of them, a merchant and philanthropist Mikhail Konstantinovich Sidorov (1823-1887), is known to be a pioneer and polar expeditions organizer. Being assure of the possibility to develop the Northeastern Sea Route, he tried to prove that the Kara Sea route was passable. To his own expense he sent ships to ascertain its ice conditions. [1]. Sidorov not only travelled but collected minerals samples from the North, sent them to exhibitions both in Russia and other countries. He also spoke about the importance of polar research in different societies. Realizing private initiative possibilities limits in such a difficult matter as the exploration of the North, Sidorov tried to include RGS in those studies.

In 1864, Sidorov offered RGS money to reward the braveman who decided to travel by sea to the Ob or Yenisei Rivers mouth. But the proposal was not supported. The RGS Vice-Chairman F.P. Litke believed that due to the difficult ice conditions in the Kara Sea, Russian sailors would not be able to sail in their vessels. Kruzenstern's failure to reach the Ob River mouth in 1862, strengthened Litke in his opinion.

However, Sidorov did not lose the hope to interest RGS in the problem of exploration the Arctic Ocean. In 1869, after discussing the polar expedition plans with the Swedish traveller A. Nordenschteld, Sidorov requested RGS "whether it recognizes to be useful to take advantage Professor Nordensköld's offer and services and individually or collectively with Sweden or Norway send to the expedition having been mentioned the Professor"[2]. The RGS Council reacted to Sidorov's idea without enthusiasm, replying that RGS had insufficient funds to organize an expedition to the Arctic Ocean.

Sidorov was not confused by the other refusal, and in March 1870 he made a report at the RGS meeting about his journey to the Pechory River mouth. Also he discussed the Norwegian voyages, which testified the Kara Sea possibility in late summer and stressed the polar expeditions importance. Having recalled the "happily initiated" the Arctic Ocean studies by Russians in Anna Ioannovna reign (referring to V. Bering's Great Northern Expedition), Sidorov asked the attendees: "Might RGS be the Siberia future main conductor by the discovery of its output directly in the sea. Might RGS provide the state benefit that the cheapest and most profitable sea route will bring [...]. Will RGO accept the petition [...] to strengthen us in our own North Sea?" [3]. Sidorov suggested that RGS should organize and send two expeditions for the Pechora and Ob Bays exploration. He believed that such expeditions would also help to study the "the New Earth continent concealing, perhaps, the untold mineral wealth". Summarizing the meeting results the Physical Geography Department Chairman P.P. Semenov noted that RGS did not posses the means to equip an expedition to the Arctic Ocean but it could draw the expedition draft.

At his suggestion, RGO set up a commission to elaborate the future expedition program. The report on the program was made by the Commission Participant P.A. Kropotkin, who took into account the different Commission members views. He included the Arctic Ocean research history, information on natural conditions and an extensive research program to the survey. Kropotkin successfully combined scarce information about the ocean flora and fauna and hypotheses about the currents and hypothetical lands distribution with new ideas about the polar region climatic conditions. The report was published in 1871 in "Izvestia RGS" journal [4] and later as a separate edition. It became the first paper in Russian geographical literature regarding the scope of the problems that the polar regions researchers had to address.

The RGS Chairman Grand Duke Konstantin Nikolaevich approved the report content and entrusted the Marine Ministry to find ways and means for carrying out the expedition. However, the idea did not meet much anticipation in the Ministry. Admiral F.M. Novosil'tsev perceived the northern expedition as Grand Duke's "sudden fantasy" (He headed not only RGS but also the Maritime Ministry) "extremely unpleasant and troublesome for the naval agency, which is though difficult to reject, at least at first" [5]. The Commission of the Naval Ministry found different reasons for avoiding the expedition organizing task. In early 1872, Konstantin Nikolayevich sent a letter to the Minister of Finance M.K. Reutern, in which he stressed that in Russian government shouldn't rely solely on private funds for studying the polar regions, like it was done in other countries, and asked whether the State Treasury to allocate 20,000 rubles for the expedition organization. The Minister of Finance replied: "Being aware of the scientific benefits of that enterprise, I can't, however, ignore that many useful and urgent initiatives are being implemented slowly or even completely postponed for lack of financing. That's why to my deepest regret, I can't agree to provide such a significant amount for the polar seas expedition "[6].

Successful foreign travellers polar expeditions in the 1870s forced the Russian society to pay more attention to the Northern challenge. Thus in 1869-1871 several Norwegian industrialists in their ships skirted Novaya Zemlya from the North and didn't find a lot of ice in the Kara Sea. They learned that at the end of summer the sea could be crossed without any difficulties. The information was very inspiring for those enthusiasts who believed that it was possible to reach the Ob and the Yenisei estuaries by sea.

In 1872-1874 K. Weiprecht and J. Payer's Austro-Hungarian expedition took place. It was financed by a large landowner G. Wilczek. The expedition ship had to go round the New Earth from the North and then reach the Bering Strait. However, it was blocked in ice to the northern side of Novaya Zemlya, drifted for a long time, and finally, in October 1873, approached the unknown land, which travellers called Franz Josef Land in honor of the Austrian Emperor.

The Austrian expedition caused a peculiar reaction of RGS. It was deplored that Franz Josef Land discovery was not due to the share of the Russian expedition. Meanwhile, Payer and Weiprecht's trip served as a pretext for Weiprecht's proposal to organize an international polar expedition.

That idea was actively supported by RGS, as well as by Academician G.I. Wild. In 1880, he headed the International Polar Commission that played an important role in organizing the Polar Year which took place in 1882-1883. In 1880, RGS organized a Commission to establish polar stations in Russia, which was headed by a physicist, RGS Member R.E. Lenz. The commission organized two stations - at the Lena mouth (on the Sagastyr Island) under the N.D. Yurgens's supervision and on Novaya Zemlya in the Malye Karmakuly village under the K.P. Andreev's supervision. The stations of all the rest countries stopped observations in 1883, while on Sagastyr Island the observations continued in 1884 [8]. The meteorological, magnetic and astronomical observations results were published in the "Proceedings of the International Polar Year" and were heavily used by scientists.

In 1885 N.D. Yurgens was awarded by RGS with Konstantinovskaya Medal for his services in the polar station at the Lena mouth. In the review for his works, R.E. Lentz stressed "The international polar station establishment seems to be undoubtedly one of the most outstanding tasks which the Imperial Russian Geographical Society enterprises for the benefit of Geography" [9]. Lentz also noted: "The scientific world owes the Society success in this matter" [10].

M.K. Sidorov's insistent aspiration to draw attention to the northeastern sea route developing didn't remain without consequences. In 1875-1876, thanks to Sidorov, A. Nordenskiöld was able to reach the Yenisei mouth by sea. The trips success inspired Nordenskiöld. He organized a new expedition with the Swedish King support. That expedition on the ship Vega took place almost along the entire Eurasia coast during one navigation. The main expedition result was the Taimyr Peninsula coastline refinement and many other various observations.

So, the north-eastern pass was first traversed due to favorable ice conditions. In 1893 the Taimyr Peninsula was safely rounded by F. Nansen. But the problem of navigation in the Arctic Ocean remained unchanged under adverse conditions. February 18, 1897 during the RGS Council meeting P.P. Semenov said that Vice-Admiral S.O. Makarov "expressed to many of the Society Members his thoughts on the navigation method in the polar seas with the steam icebreakers assistance." It was finally decided to convene an emergent RGS meeting and to propose Makarov "to explain the essence more clearly" [11].

The meeting was held at the Marble Palace on March 30, 1897. In addition to RGS Members and Ministers, the President of St. Petersburg Academy of Sciences, Grand Duke Konstantin Konstantinovich and Grand Duke Alexander Mikhailovich, heading the Naval Ministry were present. At the beginning of the meeting, Makarov's colleague the hydrograph F.F. Wrangel performed a report on the Russian polar research history. Then Makarov told about his observations results during the voyage "on the Arctic Sea" that summer. According to Makarov: "It isn't the question to build an icebreakers or not to do it, but it's the question to build them now or wait even longer" [12].

Minister of Finance S.Y. Witte supported Makarov's idea. The icebreaker Ermak was built in Newcastle (England). In 1899, when the Ermak arrived in Kronstadt, RGS sent a congratulatory telegram for its crew. But during the voyage along the Arctic Ocean to Spitzbergen, the ship's screw broke, and to fix it, it was necessary to return to Newcastle. In 1901 Makarov organized a new expedition to Novaya Zemlya.

The expedition gathered a lot of materials about the archipelago natural conditions. But the icebreaker could not overcome the ice near the Novaya Zemlya northern shore. The Ermak Arctic Ocean voyages failures resulted in a scientific community critical attitude to the icebreakers idea. The Academy of Sciences was dissatisfied with Makarov's project, since in 1899 the icebreaker could not help to take to Spitsbergen a part of the academic expedition. A skeptical attitude to the icebreaker use was also voiced in the Academic Commission on the Russian Polar Expedition Organization by E.V. Toll, (organized by the Academy of Sciences). In 1901, in a letter to S.Y. Witte, after the polar research importance discussing, P.P. Semenov firmly stated that the Ermak didn't suit for the polar voyages and, consequently, for polar studies [13].

In 1904, reporting to RGS on the tragic Makarov's death during the Russian-Japanese War, the Chairman of the Department of Physical Geography, geologist F.N. Chernyshev noted that the Ermak creation "at first seemed to be a paradox" and after the first not quite successful navigation experience provoked "criticism of the very icebreaker idea" but finally turned out to be a world discovery. Chernyshev, having himself participate in the 1901 expedition to Novaya Zemlya, organized by Makarov, believed that the Ermak was the most suitable vessel for fighting polar ice. Maritime department only after the Makarov's death realized the importance of using icebreakers for the Northern Sea Route development.

3. Conclusion

Thus, in the second half of the 19th century, the Russian Geographical Society was rarely involved in the polar research organization, partly because the polar research problem in Russia was not yet relevant, partly because of insufficient funding of RGS activities. Also there were few enthusiasts ready to complete such a difficult task. But RGS serials often published articles devoted to the polar regions observations results. Both Russian and foreign travellers' reports were promulgated at RGS meetings. Various achievements in the North study were discussed. RGS, to some extent, acquired the discussion platform role, within which polar research prospects discussion was taking place. RGS participation in research during the International Polar Year, the discussion of Nordenskiold and Nansen's expeditions results, a large number of publications on the polar regions study history and on modern polar travels attracted for a greater extent both scientists and government officials attention to the resolving the Arctic problem.

References

IOP Conf. Series: Earth and Environmental Science **180** (2018) 012001 doi:10.1088/1755-1315/180/1/012001

[1] Sidorov M K 1881 *Papers for acquaintance with the North of Siberia* (St. Petersburg: Shemetkin printing house) p 50-67

[2] Russian Geographical Society Archive F. I-1864, op. 1, d. 11, l. 44-45

[3] Russian Geographical Society Archive F. I-1871, op. 1, d 13, l. 76

[4] 1871 Report of the Commission for equipping the expedition to the northern seas *Proceedings of the Russian Geographical Society* **3** 182-184

[5] Russian Geographical Society Archive F. I-1871, op. 1, d 13, l. 76

[6] Russian Geographical Society Archive F. I-1871, op. 1, d 13, l. 106

[7] 1875 General Meeting Journal for November 5 1875 *Proceedings of the Russian Geographical Society* **6** 248-249

[8] Sukhova N G Tammiksaar E 2008 International Polar Year *International Polar Year News 2007-2008.* (St. Petersburg: Rotaprint) p 15-19

[9] 1886 Russian Geographical Society Report for 1885 (St. Petersburg: Suvorin printing house) p 3

[10] 1886 Russian Geographical Society Report for 1885 (St. Petersburg: Suvorin printing house) p 3

[11] 1897 Council Meeting Journal for February 18 *Proceedings of the Russian Geographical Society* **1** 114-115

[12] 1897 General Assembly Journal for March 30 *Proceedings of the Russian Geographical Society* **5** 401-402

[13] Vrangel' F F 1913 Excerpts from Stepan Osipovich Makarov's biography *Maritime collection* **7** 107-108

IOP Conf. Series: Earth and Environmental Science **180** (2018) 012002 doi:10.1088/1755-1315/180/1/012002

The doctrine of "common territory" versus "terra nullius": political geography in the political and legal context of Spitsbergen's status in the late 19th century – first half of the 20th century

M Zadorin

Department of International and Comparative Legal Studies, Higher School of Economics, Management and Law, Northern (Arctic) Federal University named after M.V. Lomonosov, Arkhangelsk, Russia

m.zadorin@narfu.ru

Abstract. The role of scientific and commercial development in the modern status of Spitsbergen cannot be overestimated. The archipelago development before the signing of a treaty in 1872 between Russia and the Swedish-Norwegian Union was the conflict issue undermining the "terra nullius" doctrine, which was put into circulation in the first half of the 20th century to justify the "sovereignty" of Norway on this territory in the future. At the same time, several centuries before that, the Russian Pomors, Danes and others were actively developing the archipelago, and it is confirmed by historiography and modern archaeological data. For a long time, the phenomenon of diplomatic correspondence between Russia and Sweden-Norway, known as "Agreement of 1872", did not receive sufficient coverage in scientific and legal literature. It seems extremely interesting from the standpoint of an alternative interpretation of the events related to Norwegian activation of fisheries in the Svalbard waters (this is how the archipelago is designated in the official documents of the Norwegian government) in the late 19th century – the early 20th century. The "common land" was the naming of Spitsbergen in the "Agreement of 1872". The article examines the political and legal consequences of the "terra nullius" concept and its use for the archipelago status since the signing of the Treaty in 1920, including gradual extension of the Norwegian national law to replace the international treaty regime. The expansion of the 200-mile "fishing zone" and the adoption of the "Mining Charter", which was, in fact, an intrinsic act, were the politico-geographical consequences of the "no man's land" concept, erroneously or deliberately used when signing the treaty in the early 1920s.

Introduction

Svalbard is the territory of a unique political and legal regime created due to the efforts of states to ensure international cooperation in the development of energy (primarily, coal), biological resources (fish) and security. This regime was established within the framework of the 1920 Spitsbergen Treaty (hereinafter –"Treaty"). It is also important that this act was one of the first documents in international practice that established Norway's obligations to "ensure the conservation and, if necessary, the restoration of fauna and flora". The Treaty was a kind of harbinger of the universal international standard art. 119 of the United Nations Convention on the 1982 Law of the Sea (hereinafter – "1982 Convention") related to the "reproduction" of marine bio resources. Moreover, the Convention on Fishery and

Protection of Living Resources of the High Seas, adopted in 1958 in the Art. 2 focused on "protective measures" that "must primarily ensure the supply of human food", while the 1982 Convention emphasized an "ecological approach" and protected specific species. That is why the "revolutionary character" of the Treaty had become evident at that time.

However, the "legal basis" of the Treaty, due to certain historical, political, legal and procedural specifics, has not been able to ensure full and proper compliance with provisions set out in it. This was due to the actual absence of one of the main "players", namely Russia, then the Soviet Union, among the States-parties of the agreement. The USSR joined the Treaty 15 years later (May 7, 1935), and it was the second state after Norway, maintaining a significant presence on the archipelago.

1. "Terra nullius" as a lobby tool for "Norwegianization" of Spitsbergen

The concept of Spitsbergen as "terra nullius" had a strong presence in the minds of generations and in the legal doctrine. The Treaty established the "sovereignty" of Norway over the archipelago. Art. 1 expressly set forth the Kingdom's "full and absolute" sovereignty over the archipelago within the following boundaries: from the Bear island (Bjørnøya), all islands located between 10° and 35° EL from Greenwich and between 74° and 81° NL, incl. West Spitsbergen, North-East Land, Barents Island, Edge Island, Kong Karls Land, Hope or Hopen island, and Prince Karl's land, along with all the islands, islets and rocks belonging to them.

It is believed that the concept of "terra nullius" on Svalbard was first announced by the Italian lawyer Camille Piccioni in 1909: "The issue would have been simpler if Spitzbergen, until now terra nullius, could have been attributed to a single state, for reasons of neighboring or earlier occupation. But this is not the case and several powers can, for different reasons, make their claims to this territory which still has no master." [1].

Later, systematic conferences on Spitsbergen in 1910, 1912 and 1914 did not bring any significant results in establishing a political and legal regime on the archipelago.

However, according to a researcher Mary Katherine Jones, who analyzed more than 40 Norwegian publications on the Spitsbergen issue of the 19th and early 20th centuries, a process of active "literary lobbying" of Norway's interests on sovereign issues rights over the "no-man's land" of the archipelago was on the agenda of the Paris Peace Conference of 1919 [2]. Recognition of Spitsbergen as "no-man's land" during the work of the above-mentioned conference was also promoted by a prominent American lawyer and US Secretary of State Robert Lansing [3].

Roald Berg described the policy of the Norwegian government in relation to Spitsbergen in the second half of the 19th century even more boldly: "an offensive policy of expansionism, motivated by historical and geographical considerations and alleged rights to re-establish the Medieval Norse Empire" [4]. In addition, the author also used the term "Norwegianization" with a reference to the renaming of geographical objects.

However, the history of the archipelago's development suggests an entirely different legal design, which should have been applied to the archipelago – "common areas". Professors A.N. Vylegzhanin and V.K. Zilanov underline several factors that testify to the centuries-long joint development of the archipelago by several states, incl. Britain, Netherlands, and Denmark, who, in different historical periods, claimed their territorial rights to the archipelago. Russian Pomors also owned this land but without territorial claims. Their interest was purely economic, not political. Some evidences that prove the presence of the Russian fishermen on Spitsbergen are historical facts:

1) in the 15th century, the ambassadors of Tsar Ivan III used the route to Western Europe (e.g., to Denmark in 1496) from the White Sea through the water area adjacent to Spitsbergen;

2) the active commercial development of Spitsbergen by Russian Pomors in the 15th and 16th centuries. They called the archipelago "Grumant" or "Gruland", considering it as the continuation of Greenland;

3) Pomor settlements proven by archaeological evidence based on the results of expeditions of the Institute of Archeology of the USSR Academy of Sciences in 1979–1981; for example, a house located 14 km north from Gravschen Lagoon, near Stabbelva, built in 1557;

4) international correspondence as an evidence of the Moscow possessions in Gruland:
– the letter of the Danish King Frederick II to a merchant Ludwig Munch on March 11, 1576;
– the letter from a Nuremberg doctor, Professor of philosophy and medicine Jerome Munzer to the Portuguese King Juan (Joao) II on July 14, 1493;
– the message from the Danish Admiral Severin Norby to the King of Denmark and Norway Christian in 1528;
5) early modern time mapping data:
– G. Mercator's map (1569): 7 islands were called "Svyatie Russkie" ("Holly Russians") on the site of modern Spitsbergen;
– "Karta severnykh zemel" ("Map of Northern lands"), end of 16th – beginning of 17th century: the islands were designated as "Russian land" [5].

The further active presence of Pomors on the archipelago is not contested in Norwegian historical literature. As J.P. Nielsen noted, a full-fledged Norwegian fishery began only in 1819 [6]. And we don't count scientific expeditions that allowed making significant progress in mapping and obtaining unique data. At the same time, Pomor fishermen participated in some expeditions as "local experts, sailors and pilots" [6]. Later, while negotiating the Agreement of 1872, the "common use" regime of the archipelago was confirmed. The absence of a condominium for any state provided the absolute unreasonableness of the term "terra nullius" in this regard [7].

2. Commercial expeditions as a "common use" factor

Officially, the archipelago, as well as Bear Island, located between Spitsbergen and continental Norway, is believed to be discovered in 1596 by a Dutch expedition of W. Barents. For a long time, only the coastline of the islands was investigated, because of the numerous whaling expeditions of the 17th and 18th centuries. They were operating on the coast of Spitsbergen, and did not move deeper into the island.

The first evidence of the scientific study on Spitsbergen was the expeditions of V.Ya. Chichagov in 1764–1766. They were organized upon an initiative of M.V. Lomonosov. The results of the two expeditions and their scientific observations made it possible to publish a description of the archipelago [8]. Then, English scientific expeditions of Phipps in 1773, the Norwegian Keilhau expedition in 1827, the French expedition on the ship Recherche in 1838–1839 and others were sent to Spitsbergen. Before the beginning of the 20th century, Swedish researchers played an important role in the scientific study of Spitsbergen. The most famous of them was Adolf Nordenskiöld, who compiled a detailed map of the archipelago and discovered a phosphorus-containing mineral – a raw material for obtaining artificial fertilizer on the western shores. Also, he immediately suggested that the Swedish government used the results of its research in favor of the state interests [9].

Several joint Swedish-Russian geodesic expeditions aimed at conducting the degree measurements in 1898–1902 became a notable milestone in the study of the islands.

The Russian expedition under command of the captain I rank V.Ya. Chichagov was the largest and the most complex of the late 18th century. Its goal was connected, strangely enough, with the search for the most optimal route to Kamchatka through the Arctic Ocean. According to the geographers of that time, the Ocean was ice-free. The route of the expedition was to pass along the coast of Greenland and the north-western tip of North America and then through the Bering Strait. Spitsbergen was chosen as a reference point for the start of the expedition. Two attempts in 1765 and 1766 failed, because of the ice, nebula and calm. After the mark of 80° 26' N Chichagov concluded that sailing across the Arctic Ocean was impossible. The memory of him was immortalized on Edge Island, East of Spitsbergen [6].

At the same time, the creation of the Norwegian steam fleet in the 19th century, as well as the activation of the "Norwegian State Expeditions to Spitsbergen" (Det Norske Statsunderstøttende Spitsbergenekspedisjoner – DNSS) in 1906–1925, financed by public and private donations that reached the amount of 1,875,730.03 kr. [10] by 1927, allowed Norwegians making the most effective presence and the "effective occupation" to justify their claims to sovereignty.

At the same time, it is important to note that the resources for Norwegian research in Svalbard were irregular. Until 1920, the participation of the state was insignificant. Since the signing of the Treaty it was characterized by significant growth both in absolute numbers and in relation to the state share [11].

However, Adolf Hoel, the permanent head of DNSS since 1912, highly assessed the potential of scientific activity in foreign policy: "Scientific research and exploration almost always take place prior to state annexation – political or economic – of new territories. In addition, states with the best knowledge of the region have an advantage in conflicts of interest and subsequent negotiations" [12].

3. Agreement of 1872 and the "commonly used lands" concept

In March 1871, the government of the Swedish-Norwegian Union sent a diplomatic note to several European countries, incl. Britain, France, Germany, Denmark, the Netherlands and Russia. Its content was reduced to the official intention of the government and the monarch to "take over" the Spitsbergen archipelago.

However, a further exchange of diplomatic notes involved Russia only (15/25.05.1871, 16/28.06.1872, 30.06/12.07.1872). It is known as "Agreement of 1872" in historical and politico-legal literature. Professor A.N. Vylegzhanin and V.K. Zilanov, highlighted the key theses of both sides. As for Sweden-Norway, the political and legal position was expressed as follows:

1) due to lack of settlements on the archipelago, the Kingdom citizens asked for protection and subsidies for the establishment of permanent settlements there, which would ensure their independent existence through fishing;

2) Swedish summer scientific expeditions to the islands of the archipelago had been conducted for a long period of time;

3) the islands of the archipelago had not been a part of any state, and therefore, objectively, due their nature and geographic location, they belonged to the Norwegian continent;

4) the King wished to be sure that none of the subjects of other states who had had their interests in the archipelago would oppose to the approval of his authority on that territory.

The response of the Russian diplomatic department can be described to the following position:

1) this format of interaction and consideration of interests of all interested parties was welcomed;

2) legal uncertainty was recognized regarding the rights of interested parties who at different times had tried to apply the principle of "effective occupation" of the archipelago at the expense of fisheries and permanent settlements;

3) due to the historical uncertainty, it was proposed to leave that issue without a specific political and legal solution and to apply the principle of accessibility of the archipelago for all interested parties, as before;

1) the parity of all legal "titles" of the interested parties that had ever developed the archipelago was affirmed;

2) the scientific colonization of the archipelago by Sweden based on the parity of all stakeholders was welcomed [5].

Quoting A.V. Petchurov [13] and R.V. Dekanozov [14], the authors came to a conclusion that Russia reacted negatively to the attempt to establish the Swedish-Norwegian sovereignty over the archipelago and was perfectly aware of further geopolitical risks. Addressing the diplomatic note not only to Russia, but also to other states, Sweden-Norway spoke about the established international status of the archipelago. And the absence of any protest notes regarding the statement of the Russian Ministry of Foreign Affairs on the "sovereignty" issue confirmed the consistency of the states' positions in 1872. In a closing note of June 30/July 12, 1872, the Russian government confirmed the acquaintance with the "decision taken by the Swedish Government to refuse from the annexation of the islands of Spitsbergen after consideration of the statements in the note" of the Russian Government on May 15/May 27, 1871. Moreover, customary law of the sea, associated with freedom of the high seas, shipping and fishing had been valid in the sea area of Spitsbergen until the conclusion of the Spitsbergen treatise in 1920. It is evidence in favor of the international management regime of the archipelago.

Another fundamental issue is the assessment of the "effectiveness" of the occupation of Spitsbergen by the parties to the Agreement of 1872. As L.M. Poval notes, referring to Hugo Grotius as one of the founders of the modern international public law, an effective occupation of "no man's territory" is the main condition for the international recognition of sovereignty. According to H. Grotius, the acquisition of the state territory is made either by taking possession ("occupationederelicti") or by means of a contract ("pactionibus"), or by means of conquest ("victoriae jure"). However, justifying freedom to use the sea, H. Grotius wrote that it could not be the property of any state [15]. It is obvious that at the beginning of the 1870s, there was no "effective occupation" of the archipelago by the Swedes and Norwegians and it was confirmed by the text of the Agreement of 1872.

The Agreement of 1872 reflected the foreign policy situation of the time when the Russian Empire was forced to act with the utmost caution, defending national interests. The territory of the Empire did not allow pushing its northern neighbors (especially Sweden) too actively. At that time Sweden had revanchist ideas on the return of the lost territories after the wars of the 18th century. These ideas were initiated by the Swedish King Oscar I, a successor of Karl XIV Johan (Bernadotte) [16].

One of the striking examples was a consistent change in the geography of the land border of Russia and Norway, related to the existence of the "frontier zones" phenomenon when Russia made concessions in the delimitation of the border area [17]. Moreover, on the land border of the two states, the "effective occupation" factor was of a rather controversial character, as it was linked with the infringement of rights of the Russian subjects – Lopari (pre-revolutionary Russian name of the Saami people) who led a semi-nomadic way of life [18].

This explained the actions of Russia that had the defense alliance with Britain and France of November 21, 1855 (the so-called "November Treaty") in mind and tried to use the most neutral political levers to relieve the foreign political tension in the North. This was facilitated by the failure in the Crimean War, strained relations with Britain and France, and businessmen of the Arkhangelsk unit of the Imperial Free Economic Society who did not want to incur economic losses from a possible rupture of diplomatic relations with the Scandinavians.

Conclusion

Thus, scientific and commercial expeditions to Spitsbergen were of great importance for the formation of its political and legal status. The main reasons for the loss of the Russian Empire's influence in this region are the following:

- signing of the "November Treaty" in 1855;
- the broadest geographical space of the Russian Empire and the nuances of foreign policy related to the protecting the land border priority and the "sacrifice" of the territory for economic and political purposes (the role of the "Free Economic Society" in Arkhangelsk);
- serious climate fluctuations in the Arctic region [19], which influenced the formation of year-round fishing in Norway due to the Gulf Stream, and limited fishing in Russia, only along the coast in the warm season (May – October);
- Norwegian powerful steam fleet and their scientific and commercial development as an "effective occupation" factor;
- actual absence of a full-fledged merchant navy in the Russian Empire up to the Soviet period;
- removal of Russia as a political "player" in the negotiations on the Svalbard Treaty in 1920.

After the Svalbard Treaty of 1920, which implied a clear legal ambiguity and uncertainty about sovereignty and at the same time the international legal regime, the Norwegian government began to use various legal subtleties for the gradual extension of national law to the archipelago.

As it was mentioned by the lawyer and legal expert of the Arctic A.M. Oreshenkov, Norway managed to overcome Art. 2 and Art. 7 of the Treaty. The articles secured that any person who occupied a part of the land, upon fulfillment of the procedure established by the Treaty, got recognized ownership of it, including the rights to mining. The Norwegian part, using Art. 8, which established the requirement for the adoption of the Mining Regulations for Spitsbergen, developed an agreement not on the principles of the Treaty, but as a national normative legal act, according to which individuals and companies did not acquire the right of ownership of land but right of urgent use of mining outlets. In A.M. Oreshenkov's opinion, this meant that the persons who occupied the land before the contract lost some of the acquired rights, namely, the rights to mining. The Norwegians used the conflict created between the principles of the Treaty and the mining charter in their own interests. The Norwegian documents on the exclusive right of occupied land ownership provided the recognition of neither acquired rights (violation of Art. 6 of the Treaty) nor exclusive property rights to them (violation of § 9, § 1 and § 11 of the Annex to the Treaty). In addition, the Norwegian government extended the scope of the "Mining Charter" to the territorial waters of the archipelago, including the continental shelf, which was also a violation of the Treaty regime of the archipelago.

Also, in the late 1970s, Norway established a 200-mile fishing zone, considering the provisions of the UN Convention on the 1982 Law of the Sea, which extended to the Treaty regime of the archipelago, without any regard to the fact that the Treaty itself did not provide the existence of such a zone.

Thus, it becomes obvious that the marking of Svalbard as "terra nullius" until the signing of the Treaty of 1920 existing to this day in a large layer of historical, political and legal literature, is the starting point for the violation of the international common use regime in relation to the Spitsbergen archipelago.

In conclusion, it cannot but be mentioned a scientist Adam Grydehøj who believes that Norway's Svalbard policy is historically focused on marginalizing Russian influence; but at the same time, however, it risks losing authority within Svalbard due to the strengthening of local democracy in Longyearbyen and increasing opportunities for the involvement of non-traditional Arctic actors such as the Asian economic powers [20].

Gratitude
The article was supported by the Presidential science foundation (project MK-6382.2018.6 "Reconstruction of the symbolic and political boundaries of the sea and island areas of the Western Sector of the Russian Arctic in the 19th – early 20th centuries").

References

[1] Piccioni P 1909 Revue generale de droit international public **XVI**
[2] Jones M K 2014 Spitsbergen Literature Lobby *Nordlit* **32** 33–69
[3] Polar Politics: Creating International Environmental Regimes 1993 (Cornell University Press) 85–87
[4] Berg R 2013 From "Spitsbergen" to "Svalbard". Norwegianization in Norway and in the "Norwegian Sea", 1820–1925 *Acta Borealia. A Nordic Journal of Circumpolar Societies* **30** (2) 154–173
[5] Vylegzhanin A N, Zilanov V K 2006 Spitsbergen: Legal Regime of Adjacent Marine Areas (Moscow: SOPS Publ) (in Russian)
[6] Nielsen J P, Mankova P 2017 Russland kommer nærmere: Norge og Russland 1814–1917 (Moscow, Ves' Mir Publ) (in Russian)
[7] Oreshenkov A M 1992 Features of the international legal regime of Svalbard *Moscow Journal of International Law* **3** 128–135 (in Russian)
[8] Arlov T A 1994 short history of Svalbard (Oslo) 43–44
[9] Kasiyan A S 2013 The Spitsbergen Issue in the Russian-Norwegian Relations (1870s–1953) *Journal "Vestnik of Northern (Arctic) Federal University" Humanitarian and Social Sciences* **1** 16–22 (in Russian)
[10] Saburov A A 2015 The precursors of the Norwegian Polar Institute: "Norwegian State Expeditions to Spitsbergen" and "Norwegian Service for the Study of Spitsbergen and Polar Regions". Features of financing in 1906–1940 *Uchenye zapiski kafedry vseobshchey istorii* **3** 6–24 (in Russian)
[11] Saburov A A 2015 Resource support for Norwegian research on Spitsbergen in 1906–1925

Sotrudnichestvo i sopernichestvo v tsirkumpolyarnom regione: istoriya i sovremennost'. Sbornik nauchnykh statey studentov, aspirantov i molodykh uchenykh 32–38 (in Russian)

[12] Saburov A A 2016 Planning of Norwegian scientific activities on the islands of the European part of the Arctic in 1906–1925 *Sotrudnichestvo i sopernichestvo v tsirkumpolyarnom regione: istoriya i sovremennost'. Sbornik nauchnykh statey studentov, aspirantov i molodykh uchenykh* 78–84 (in Russian)

[13] Pechurov A V 1983 Shpitsbergen (Moscow:Yurist Publ)

[14] Dekanozov R V 1966 The international legal position of Svalbard: PhD thesis (in Russian)

[15] Poval L M 2012 Russian-Norwegian agreements on the division of Arctic spaces *Arctic and North* **6** 118–144 (in Russian)

[16] Musaev V I 2014 Russian-Scandinavian relations and Russian interests in the North of Europe (second half of the 20th – early 20th century) *Skandinavskie chteniya 2012 goda: Etnograficheskie i kul'turno-istoricheskie aspekty* 158–170 (in Russian)

[17] Goldin V I , Zaikov K S , Tamitskiy A M 2015 The Role of the XVIII Century Maps and the Population of the Higher North of Russia and Norway in the Construction of the Borders of the Internal Sovereignty *Bylye Gody* **38** (4) 849–864

[18] Goldin V I, Zaikov K S, Tamitskiy A M 2015 Sami in the History of the Norwegian-Russian Borderland: Factor of Tension or Regional Integration? *Bylye Gody* **37** (3) 519–535

[19] Klimenko V V, Astrina N A 2006 Documentary evidence of strong climate variations in the Russian Arctic in the 15th–20th centuries *Istoriya i sovremennost'* **1** 179–217 (in Russian)

[20] Grydehøj A 2014 Informal diplomacy in Norway's Svalbard policy: the intersection of local community development and Arctic international relations *Global Change, Peace & Security* **26** (1) 41–54

Intellectual potential for the Arctic ecosystem development

V V Glukhov[1], **I L Tukkel**[2] **and G F Detter**[3]

[1] Peter the Great St. Petersburg Polytechnic University, Doctor of Economic Sciences, Professor, St. Petersburg, Russia
[2] Peter the Great St. Petersburg Polytechnic University, Doctor of Technical Sciences, Professor, St. Petersburg, Russia
[3] Arctic Research Center of the Yamal-Nenets Autonomous District, Candidate of Economic Sciences, Salekhard, Russia

detter@mail.ru

Abstract. The level of the society human capital development in new technological order becomes the determining parameter of successful development. Glocalization processes are growing against the background of globalization, ensuring the integration of regional economies into technological and financial chains. The development of the continental and marine Arctic is accelerating. The most effective and environmentally safe development will take place in technologically advanced ecosystems that have created a favorable institutional environment. Both these factors depend on the level of human capital development. Only the country (region), which makes efforts for the purposeful formation of human capital, has the opportunity to steadily develop the Arctic and benefit from it. In connection with climatic, conjuncture and technological challenges, the Arctic zone of the Russian Federation will face serious challenges in the coming years. Responding to challenges of the XXI century, rapid and effective mobilization of scientific and technological initiatives accompanied with the search for talent and active entrepreneurs are required. Regional personnel initiative will raise the level of human capital, which is especially important in the Arctic zone of the Russian Federation. It should be noted that not a single Arctic region of Russia currently possesses technological capabilities, scientific and research potential, entrepreneurial initiative and finances necessary for a scientific and technological breakthrough. In rare cases some are able to offer sustainable, advanced and demanded scientific-technological and technical solutions. The proposed regional personnel initiative is aimed at outstripping the increase in the general (regional and attracted) intellectual potential that will ensure the development of Arctic ecosystem.

1. Introduction

Exploration and extraction of minerals in the Arctic and related issues of ecology and environmental protection pose serious challenges for the Arctic states. A proportional response to them can be given by countries and regions that attach great importance to the education and training of highly qualified personnel. The level of development of human capital is essential for the exploration and development of the Arctic. The Arctic was developed by the efforts of many researchers and pioneers, who created opportunities for the founding of industry and new cities. Northern countries and regions are largely indebted to people who have studied the Arctic areas, since a significant part of their wealth is still being formed in these territories. «Human capital was highly valued and constituted an object of national pride, but studies of this phenomenon did not receive serious development. In the planned economy, human capital was considered primarily as an input factor of production and, therefore, with the advent of market relations, became the subject of optimization for new owners, especially in the Arctic territories» [1].

2. Main part

The main goal of the Arctic development (or redevelopment) [2] is determined by the need to replenish Russia's strategic resource base in order to meet the challenges of the country's socio-economic development [1]. Within this framework, the goals and objectives aimed at the integrated security of the region, the preservation of the ecological balance, the observance of indigenous peoples' interests, and the development of the science of technology as a tool for achieving the main goal are set as secondary tasks.

Scientific and technological support of the purposes of the Arctic development is carried out through existing and formed innovative systems. Studies of the theoretical and methodological foundations and practical results of the innovative ecosystems formation in the Arctic make it possible to assert [3] "that in all strategies, in their movements and interlocks, reliance on a system that initiates and promotes innovation, on a regional innovation system will meet today's and tomorrow's challenges, the objective and global development of the sixth wave of innovation, the emergence of a 'knowledge economy'".

Innovative development is based on fundamental and applied scientific knowledge [4]. Traditional knowledge can also form an innovation process [5]. However, studies conducted in the circumpolar region, are often "about these places", and "not for these places", the North is regarded as a "laboratory for scientific research" [6]. However, if development in the Arctic should be more sustainable and human-centered, it is necessary to change approaches [7]. This same point of view is also expressed by Russian researchers, in particular: "The Arctic zone of Russia is a large site of development of fundamental and applied scientific research in various fields of knowledge ..., Russian Arctic cities can become intellectual platforms of a new economy based on knowledge and innovations" [8]. Therefore, the future opportunities of Russia's Arctic regions will largely depend on scientific and technological initiatives that will allow getting benefit from new technologies in the interests of business and society. This is also relevant for Western Arctic countries.

The Arctic zone of the Russian Federation (Russian Arctic) fully or partly includes nine subjects of the Russian Federation, or, more precisely, 271 municipalities. The area of the Arctic territories of Russia is 3.7 million km^2, the population is 2 million 420 thousand people, while the share of GRP is 5.3% of the whole Russian GDP. The population of the Arctic regions of the Russian Federation has decreased significantly since the 1990s, especially in the northern territories. At present, some decrease in population in several regions continues. In regions with a more developed economy there is a stabilization or a slight increase in the population. The share of high-tech and knowledge-intensive industries in the GRP of the macroregion is 7.1%, while the average for Russia is 20.7%. The share of high technology innovative goods in the total volume of shipped goods is 0.13%, which is 10 times lower than the Russian one.

The potential of the scientific and educational system of the Arctic regions is represented by 18 higher educational institutions, 12 of which are branches of higher educational institutions, 9200 specialists annually graduate from. There are 62 scientific organizations in the territory of the Russian Arctic, where 1900 researchers work. Almost the whole scientific and educational potential of the Russian Arctic is concentrated in the cities of Apatity, Murmansk, Arkhangelsk, in an insignificant part – in Naryan-Mar, Salekhard, Nadym, Norilsk and Anadyr. The point placement of the system elements, creating knowledge and competence, forms a high differentiation of human capital in the Russian Arctic and the need for its development.

"Human capital is a set of innate abilities and acquired knowledge, skills and motivations, the effective use of which contributes to the increase of income and other benefits. Analysis of the real processes of development and growth in modern conditions led to the affirmation of human capital as the main productive and social factor in the development of modern economy and society" [9]. In many ways, it is formed in the systems of university and postgraduate education.

According to the results of the study of the higher education accessibility [10] in the regions, the Arctic regions showed the worst result, having settled in the last third of the list, with the exception of the Krasnoyarsk region (Table 1).

Table 1. List of regional higher education systems in the Russian Arctic

Arctic Region	Number of places	Population 17 - 25 years	Coverage, %	Place
Krasnoyarsk region	96 115	324 794	29,6	36
Murmansk region	14 114	80 508	25,3	57
Arkhangelsk region	22 777	108 872	23,6	68
The Republic of Sakha (Yakutia)	30 422	124 869	24,4	63–64
Yamal-Nenets Autonomous District	1957	53 121	10,5	80
Chukotka Autonomous District	181	4 690	10,2	81
Nenets Autonomous District	0	4 320	0,0	82

The above data reflects the situation at the end of 2016, yet a decade ago the situation was much better. For example, in 2007 in the Yamal-Nenets Autonomous District 13440 people were enrolled in higher education institutions [11]. According to monitoring results in 2017, there were 2 branches of a higher educational institution, the number of places was reduced to 1957. At the same time, the number of people aged from 17 to 25 years is more than 50 thousand. Young people who want to get a higher education, are forced to leave to study in Tyumen, Moscow, St. Petersburg, and as a rule they do not returnafter the study. The absence of universities and other research organizations in the circumpolar regions is seen as a factor hampering innovative development [14].

"In the 70s - 90s of the XX century, European Arctic countries created regional higher educational institutions with a wide range of applied specialties. Having created an effective system of training scientific and engineering personnel and providing enterprises with them, they were able to make a breakthrough in innovative development and ensure accelerated economic growth" [12].

At present, the development of intellectual capital through the creation of classical universities in the Arctic cities has no prospects, since it is expensive both in capital expenditures and in operating expenses associated with the formation and retention of a talented team of research and teaching staff. It seems more promising to consider evolutionary development of vocational education institutions in the Arctic regions, the qualitative improvement of their level on the basis of network, remote and online technologies, the development of a research base for the formation of colleges of new generation "College 3.0", which presupposes the succession of improving competencies through education - research - implementation. Organization of regional innovation and technology centers on the basis of colleges in cooperation with small and medium-sized businesses will form a favorable innovative ecosystem and will allow benefiting from new technologies in the interests of business and society.

Undeveloped intellectual infrastructure of the Arctic regions provokes population flows. On the one hand, young people who have lost touch with the small homeland prefer to stay in big cities even having problems with employment, on the other hand, specialists from the central and southern regions come to the Arctic region on a temporary basis, they are not adapted to local climatic conditions and this entails negative social, environmental and economic consequences for the recipient region. At the same time, transaction costs associated with moving and living in a new place increase for both categories of the population (migrants and immigrants). To date, no studies have been conducted on the costs incurred by organizations to attract specialists from other regions and the level of qualifications of these specialists. The question remains whether the attracted specialists will become patriots of the region, whether they will work with full efficiency for its development, or they will be guided only by economic interests. A significant negative consequence of the fact that young people leave their small homeland is the subsequent departure of their parents, which leads to the removal of both human and property capital accumulated during all years of life in the North, thus reducing the economic potential of the region.

Training the endemic population in major institutions of higher education in well-developed scientific, educational and industrial centers does not solve the problem of young specialists' return to their native cities either. It is necessary to solve not only the problem of where and how the schools graduates will continue their education, but also where and in what industry they will be able to find a job after the training. Education should be based on the principles of targeting orientation, which presupposes the preparation of formed groups of graduates in secondary and higher vocational schools, focused on the actual demand of specific enterprises and institutions in the region (city). Thus, the students know where they will work after the successful completion of their studies, the HR departments know who will join the organization in the future. At the same time, if it is expedient to return graduates of educational institutions to work to their native homeland, it is necessary to form not only an industry (organizational) procedure for the personnel training, but also a territorial one for specific settlements and professions. Taking into account the modern development of information technologies, the solution of such a task is quite manageable.

The fixed decline in the indicators of social development of the Arctic regions accompanied with industrial development shows that in the emerging knowledge economy even successful regions that did not create their own highly qualified personnel training system expect a decline in the level of human capital and subsequently the loss of leadership in the process of upcoming technological transformations.

This becomes possible, because in the conditions of a new technological order formation, the development of regional economy can be ensured only by embedding regional business in new technological and financial chains. Inclusion of the region in financial and technological flows is possible after solving two problems: the formation of a favorable institutional environment; improving the technological effectiveness of the economic system. The determining parameter for solving these problems is the level of development of human assets in the regional community. Thus, only the region that makes efforts for the purposeful formation of human capital has the opportunity to gain a foothold in the markets of the future.

Considering a person, his/her abilities and resources as the main capital of the Arctic regions, a regional personnel initiative that involves the training of highly qualified personnel in the interests of the region, the endemic population and business, should play a key role in the harmonious development of the Arctic territories and the use of the Arctic riches.

3. Conclusions

The proposed regional personnel initiative is aimed at building the human capital of the region through the formation of highly qualified personnel on the basis of endemic population and the reproduction of labor resources corresponding to the needs of the region in terms of professional composition. It is necessary to solve the following priority problems:

- development of integrative scientific and educational links with institutes of the Russian Academy of Sciences, universities, scientific centers;
- formation of favorable professional and social conditions for the educational and research activities of the population;
- assistance provided to graduates of schools from regional authorities and employers in further training and employment; development of human resources innovation infrastructure in the region;

- formation of colleges of new generation "College 3.0", which presupposes the succession of improving competencies through education - research - implementation, organization of regional innovation and technology centers on their basis in cooperation with small and medium-sized businesses.

Particular attention should be paid to the advanced development of the personnel subsystem of the innovation infrastructure. The main personnel infrastructure subdivisions - personnel centers are formed in cities as part of the local innovation infrastructure and are consolidated in the regional personnel center, which provides the forecast of staffing requirements from the perspective of the long-term scientific and technological strategy; training of representatives of government and business in the areas of innovation theory and innovation management for the development and implementation of innovative projects.

References

[1] Detter G 2017 Substantiality of Nature-Friendly Technologies in the Sustainable Development of Arctic Regions of Russia *Modern Scientific Thought* **1** 237 (in Russian)
[2] Alekseev G, Antipov S, Afonichkin A, Afonichkina E and Baburov S 2016 Arctic space of Russia in the XXI century: development factors, organization of management (Moscow: Nauka) p. 1016 (in Russian)
[3] Glukhov V, Detter G and Tukkel I 2015 Creation of a regional innovation system in the Arctic zone of the Russian Federation: design and implementation experience *Innovations* **5** 86 (in Russian)
[4] Asheim B and Gertler 2005 Regional innovation systems and the geographical foundations of innovation *The Oxford Handbook of Innovation* 291
[5] Asheim B, Coenen L and Vang J 2007 Face-to-face, buzz and knowledge bases: Sociospatial implications for learning, innovation and innovation policy *Environment and Planning C: Government and Policy* **25** 655
[6] Heinen L 2005 Impacts of globalization, and the circumpolar north in world politics *Polar Geography* **29** Available online 2011 htt ps://doi.org/10.1080/789610127
[7] Kaul I 2002 Challenges of human development in the Arctic. (New York: United Nations Development Programme) Available from: http://www.arcticparl.org/files/static/conf5kaul.pdf [Accessed 15th April 2018]
[8] Pilyasov A 2009 And the last will be the first: the northern periphery on the way to the knowledge economy *M.: Librokom* 544 (in Russian)
[9] Becker G 1993 Human Capital: A Theoretical and Empirical Analysis with Special *The University of Chicago Press* 123
[10] Gromov A,Platonova D, Semenov D Pyrova T 2016 Accessibility of higher education in the regions of Russia *Modern Analytics of Education* **8** 32 (in Russian)
[11] Trippl M, Zukauskaite E , Healy A and Marques P 2016 Smart specialization for regional innovation work *Smart Specialization Working Paper Cardiff University* Package 5 Deliverable 5.2
[12] Detter G 2015 Genesis of National and Regional Innovation Systems of the Arctic States in the Context of National Innovation Policies *Scientific Bulletin of the Yamal-Nenets Autonomous District* **3** 15 (in Russian)

IOP Conf. Series: Earth and Environmental Science **180** (2018) 012004 doi:10.1088/1755-1315/180/1/012004

Realizing clean technology principles for the mining and processing industry in the Arctic region

M Vasekha[1], V Pokholchenko[1], Zh Vasilieva[1] and M Chislov[2]

[1] Murmansk State Technical University, Murmansk, Russia
[2] St. Petersburg State University, St. Peterburg, Russia

vasyoha@mail.ru

Abstract. The production of apatite concentrate in the Arctic region of the Russian Federation is carried out using technology developed during the middle of the twentieth century. One of the most energy-consuming and environmentally dangerous stages of this technology is the drying of raw concentrate. Energy for drying is obtained from combustion of fuel oil, which leads to greenhouse gas emissions and environmental pollution. This study analyses the drying step in the apatite production process to optimize the environmental and economic aspects of this process. The characteristics of apatite concentrate during heating was studied through synchronous thermal analysis. Mass loss was determined during each stage. It was observed that drying of apatite concentrate does not require high temperatures, which are achieved in drum chambersthrough fuel oil combustion. Therefore, drying with flue gases is inexpedient. A method that involves dewatering the concentrate, which can be considered a form of clean technology and does not involve fuel oil combustion, is proposed. The results of this study can be used to modernize the current technology for the production of apatite concentrate.

1. Introduction

The Arctic region of the Russian Federation is characterized by high susceptibility to anthropogenic impact and low regenerative properties. Technogenic and human impact factors influence mining and processing of minerals and raw hydrocarbon materials. The environmental management systemin some areas of the Russian Arctic has generated highly modified territories [1]. Natural geochemical conditions, exhaustion of biodiversity, soils and vegetation degradation, erosion, and environment pollution have been observed. Therefore, optimization of economic and ecological balance is an important goal.

In these conditions, the urgent task of developing Arctic regions requires striking a balance between economy and ecology. The world is presently transitioning from the traditional model of economic growth to a «green economy» based on resource conservation and clean technologies [2, 3]. Thus, the key task of creating a green economy in the Arctic is the transition to clean processing technologies.

This task is especially important for the mining and mineral processing industries in the Arctic region. The primary strategy for development of promising mining and processing complexes should focus on optimization of technological processes while reducing environmental impact, which corresponds to the principles of the concept of clean technology.

An example of the transformationof production processes towards cleaner technologies is the modification of the apatite concentrate technology in JSC Apatit (Murmansk region, Russia).

2. Analysis of the apatite concentrate technology

JSC Apatit is the world's largest apatite concentrate production enterprise, which develops deposits of apatite-nepheline from endogenous (magmatic) origin ores in the Khibiny Mountain Group [4].The high-grade apatite concentrate contains 39 % or more of phosphorus pentoxide (P_2O_5), which is used as the feedstock for phosphate fertilizers and production of feed phosphates [5]. The concentrate, also contains 32.7±3.1 % calcium, 0.57±0.13 % silicium, 0.41 % aluminium, and traces of nonferrous and rare-earth metals [6]. Production output reached 9.5 million tons per year [7].

Figure 1 shows the basic technological scheme of apatite concentrate production. The technological scheme includes the following basic operations: splitting the initial ore, ore grinding, flotation separation of apatite with phosphorus (V) oxide content of not less than 39 %, concentration in thickeners and hydrocyclones, filtration with separation of the filtrate and cake, and final drying of the apatite concentrate.

Figure 1. Basic technological scheme of apatite concentrate production.

Apatite concentrate is traditionally dried in rotary-type drum chambers by convection. The drying agent is air heated by combustion gases from fuel oil combustion. The main disadvantage of the traditionally-used rotary chambers is their low efficiency and very high energy consumption. Energy consumption during drying can be reduced using more modern methods.

The use of rotary-type drum chambers has many indirect disadvantages. Condensate formation leads to destruction of buildings during humidification and due to freezing in winter, metal corrosion is activated, and solid micro-particle and convection gas emission during evaporation and drum rotation.

The drying chamber furnaces use M-100 brand fuel oil (high-sulfur), and consumption reaches 10.34 kg per ton of apatite concentrate. Thus, the annual consumption of fuel oil is 97.8 thousand tons, which is a relatively large amount of fuel resources and energy.

The necessity of reducing fuel oil consumption during apatite concentrate drying is obvious. This is being pursued both in the field of resource and energy savings, and directly reduces environmental impacts.

This study analyses the apatite concentrate drying process to improve and optimize the environmental and economic aspects of the production process. A synchronous thermal analysis was carried out using the Netzsch STA 449 F1 Jupiter complex with a Netzsch QMS 403C Aëolos quadrupole mass spectrometer.

Apatite concentrate was obtained at the filtration stage from the apatite-nepheline enrichment factory (JSC Apatit, Murmansk region, Russia) and was used for analysis.

Modelling of apatite concentrate drying was carried out at 30-1050 °C and under argon flow (100 ml/min). Drying was performed in a platinum crucible. The apatite concentrate drying results are presented in Figures 2-6.

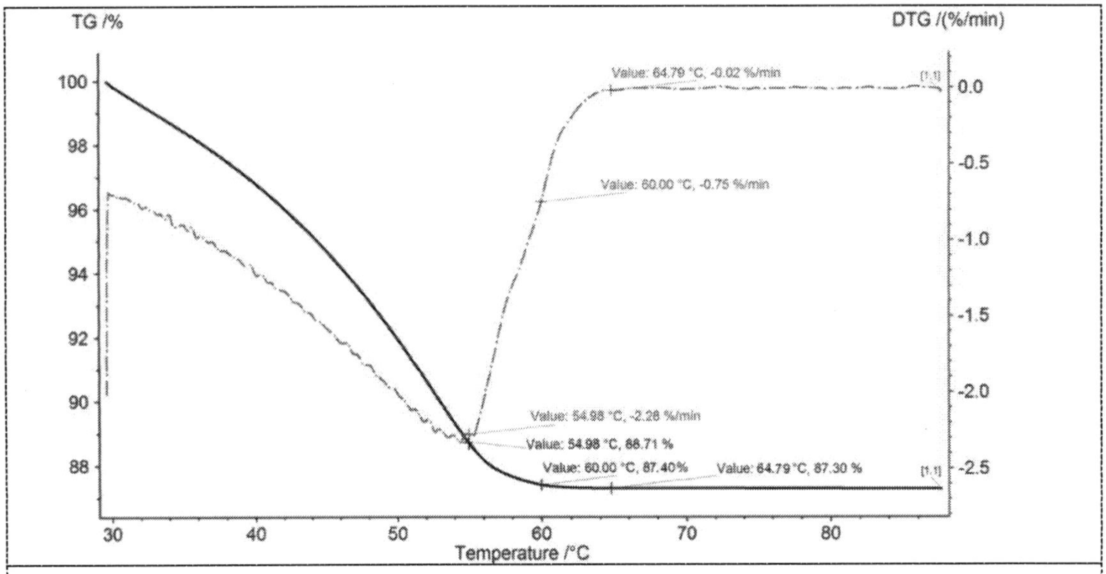

Figure 2. Apatite concentrate moisture content change in argon at various temperatures.

The dependence presented in Figure 2 shows that dehydration of apatite concentrate ends at 60 °C. After this value, the TG-curve runs parallel to the x-axis. This means that drying in the hightemperature regime, which is achieved by fuel oil combustion in air, is not required.The maximum drying rate is observed at 55 °C (downward peak on the DTG-curve). Figure 3 shows the specific heat of the apatite cake within the temperature range presented in Figure2.

The differential-scanning calorimetry (DSC) signal shows how heat flow is related to the material's mass. The heat flow is related to a specific phase transition heat and indicates how the material loses weight per unit time. This allows the temperature at which the maximum dehydration rate is observed to be determining. This makes it possible to select the optimal drying parameters.

The DSCsignal of apatite concentrate drying indicates an endothermic effect within the investigated temperature range. The latent heat at the phase transition is 2143 J/g. This is slightly below the 2300 J/g enthalpy value of free moisture evaporation.

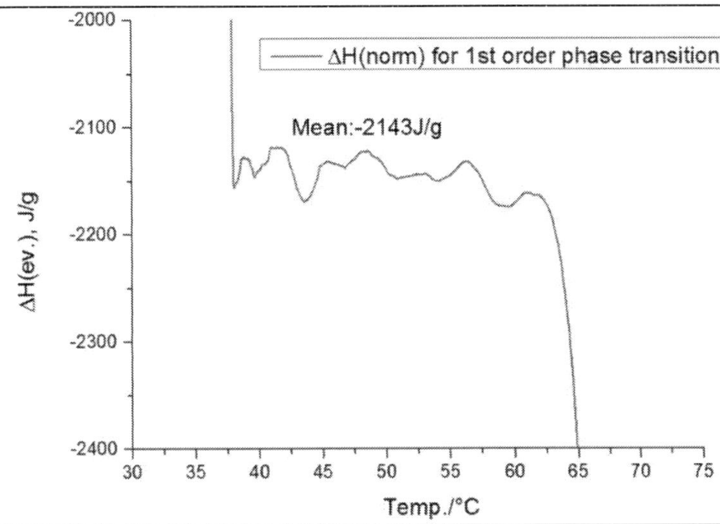

Figure 3. Phase transition energy of apatite raw concentrate obtained at the filtration stage during. The results were obtained using differential-scanning calorimetry.

We can suppose that decreased drying enthalpy of apatite concentrate compared to pure water evaporation is due to the influence of surfactants contained in the raw apatite cake obtained after filtration.To prove the supposition that high temperatures do not cause significant changes in apatite concentrate composition, thermo-gravimetric analysis was performed at temperatures up to 1050 °C. These results are shown in Figure4.

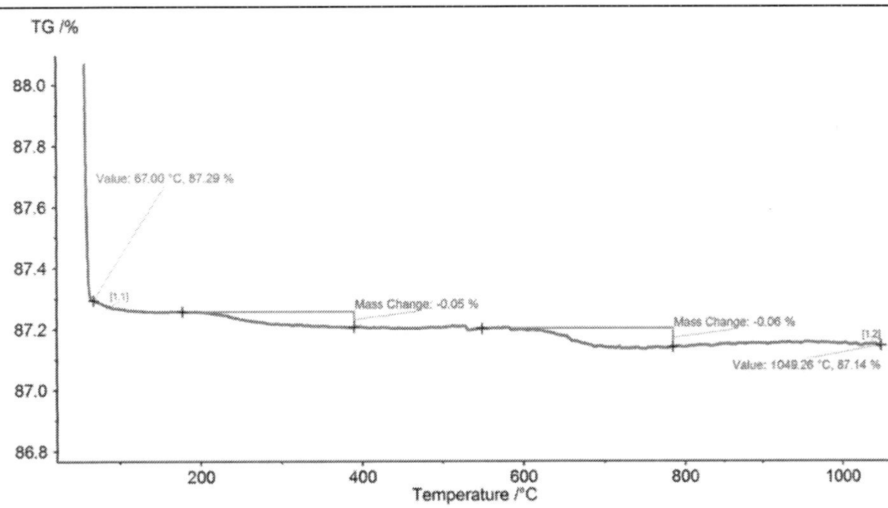

Figure 4. Thermogravimetric analysis of the apatite concentrate during drying.

One can see that subsequent loss of mass at temperatures ranging from 67 to 1050 °C proceeds in two stages. The first stage proceeds at 212 °C, and the second at 757 °C.

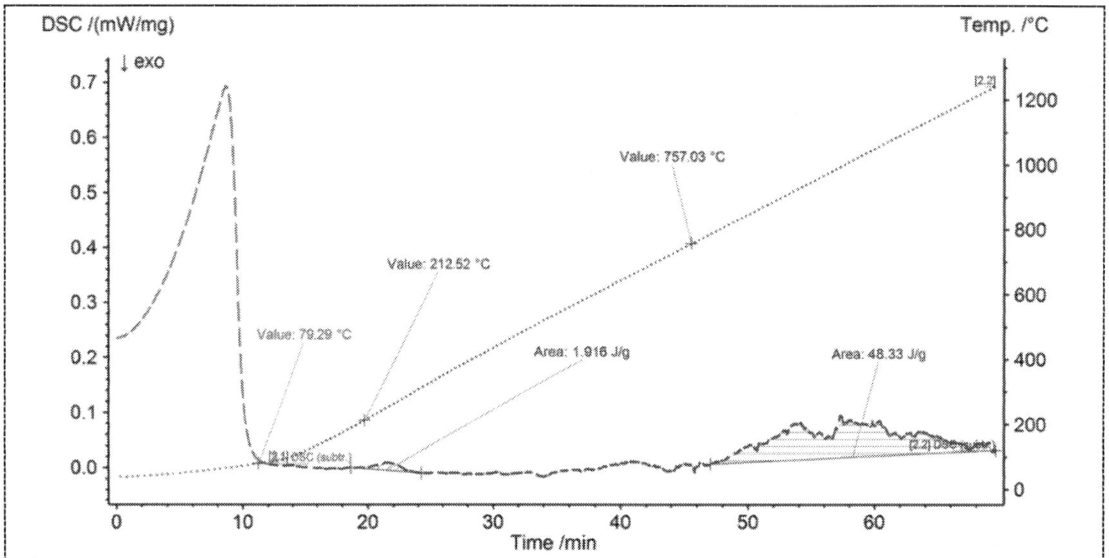

Figure 5. Endothermic effects at 212 °C and 757 °C.

The endothermic effects occurring in the apatite concentrate are shown in Figure 5. From this figure, it can be observed that there are two endothermic peaks at 212 °C and 757°C. The total mass loss is insignificant and is about 0.1%.

During the first stage of heating at 212 °C, crystallized water is released, amounting to 0.05 % of the total sample weight. The obtained results confirm the data from similar studies carried out with respect to Moroccan phosphate, which is an analogue of apatite concentrate [7]. In this case, studies have shown the presence of an endothermic peak due to moisture removal in the 100-200 °C temperature range and significant endothermic carbonate peaks at 795-815 °C and 830-890 °C. In contrast to the studies under discussion, two exothermic peaks at 300-400 °C and 730-770 °C were also observed, corresponding to the combustion of organic impurities. Apatite concentrate (Kola Peninsula, Russia) does not contain organic impurities. Therefore, heating up to high temperatures for burning organic matter is not required.

During the second stage at 757 °C, weight loss refers to the release of carbon dioxide (Figure 6). It can be assumed that CO_2 is released due to decomposition of carbonate impurities. Phosphorus and fluorine in the gaseous products formed upon heating to high temperatures were not found.

Thus, the analysis of prior studies has shown that high-temperature drying of apatite concentrate, currently being implemented at JSC Apatit, is impractical, in view of the fact that dehydration ends at 70 °C. In addition, there are no organic matter impurities in the apatite concentrate (Murmansk region, Russia), and this material does not require hightemperature processing. However, in the technology used at JSC Apatit, the temperature of flue gases entering the rotary kiln ranges from 600 to 1000 °C.

An energy calculation showed that about 327 MJ of energy is required to dry 1 ton of crude apatite concentrate. In this case10.3 kg of M-100 grade fuel oil are consumed during drying of 1 ton of raw concentrate, which leads to 418 MJ of energy consumption. Thus, when drying each ton of raw apatite concentrate cake, about 91 MJ are lost to the environment, which is about a quarter of the total heat energy produced by fuel oil combustion.

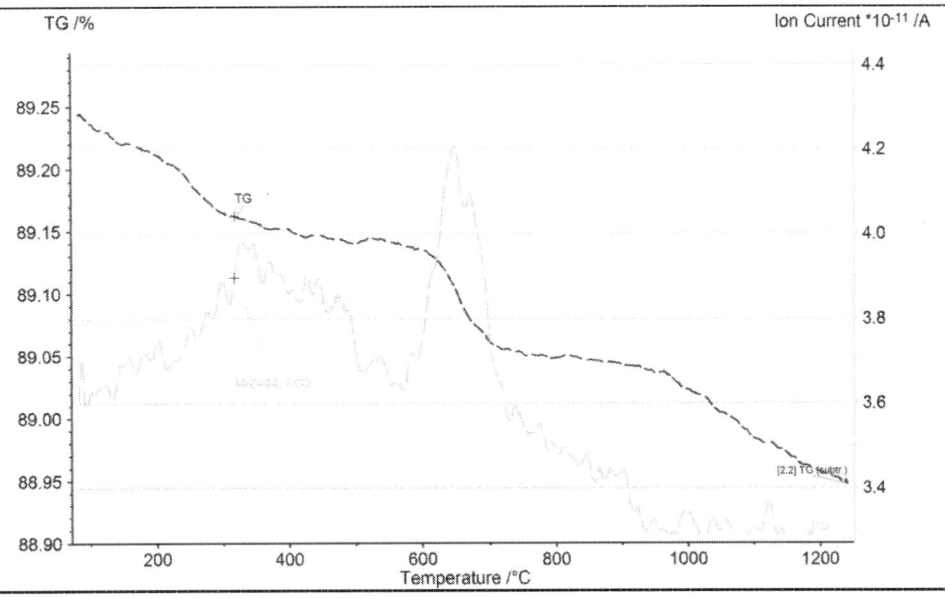

Figure 6. Mass spectrometric determination of carbon dioxide.

The calculated environmental load showed that in addition to the low energy efficiency of the drying process, fuel oil combustion causes significant ecological damage to the ecosystems in the Arctic region. Combustion of high sulfur fuel oils results in the release of greenhouse gases, black carbon, significant amounts of nitrogen and sulfur dioxides, and carcinogenic substances, including polycyclic aromatic hydrocarbons and other toxic compounds [8, 9]. The emission levels and oxygen consumption during heavy fuel oil combustion are presented in Table 1.

Table 1. Specific emissions and oxygen consumption during combustion of high sulfur fuel oil.

M-100 high sulfur fuel	Consumed during combustion, t		Emissions from fuel oil combustion, t			
	Oxygen mass	Air mass	Carbon dioxide	Sulphur dioxide	Nitrogen oxides	Particulate matter (PM)
1 ton	3,1546	13,6327	3,2296	0,07	10,4831	0,00092

The use of clean technology principles requires the introduction of new solutions for the implementation of energyefficient and environmentally friendly drying technology for apatite concentrate. New apatite concentrate drying technology should consider the following points, justified by the above research results. The greatest energy costs during drying of raw material are noted during the phase transition of liquid water to steam. Thus, clean technology should include drying apatite concentrate without phase transition «water-vapour». This approach will save a significant amount of energy and reduce greenhouse gas and toxic gas emissions caused by fuel oil combustion.

One of the possible technological solutions in this direction is ultrasonic drying. Ultrasonic treatment allows moisture removal in the form of fog (microdroplets) rather than steam [10], which can significantly reduce energy consumption. At the same time, there is no need for energy consumption for heating of heat carriers, elements, and structures in the drying chamber.

To test the effectiveness of ultrasonic drying, laboratory studies were carried out using a laboratory ultrasonic generator with an output of 0.07 kWh. The productivity of the laboratory unit was 2.08 kg/h of apatite concentrate. In this case, the energy consumption per 1 ton of raw cake will be 121 MJ, which is comparable in energy efficiency to 3 kg of burnt fuel oil. The residual moisture index following ultrasonic drying of the concentrate was 1± 0.05 %.

An alternative solution may involve using a heat pump system in a water-to-air system, for example, and using a factory recycled water supply as a low-potential heat source. According to the available data, and using JSC Apatit as an example, the circulating water supply amounts to about 70 l/min with a water temperature of 10 to 15°C, which is sufficient to operate the heat pump system in an efficient mode with a heat conversion coefficient of 3-3.5.

3. Conclusion

Analysis of apatite concentrate production showed that one of the most energy-consuming and environmentally hazardous operations is the drying of raw concentrate. This process was studied using synchronous thermal analysis. The results show that the dewatering of the concentrate ends at a temperature of 60 °C. With further heating, the mass loss is about 0.1% due to water evaporation and release of carbon dioxide due to decomposition of carbonates. The absence of organic impurities in the apatite concentrate means that high-temperature heating is not required for combustion. It is economically and environmentally expedient to conduct drying without heating to high temperatures. In such conditions, a cleaner water removal technology is possible. The authors have tested the method of ultrasonic drying, which allows one to lower the moisture content of the concentrate to a commercially viable value with much lower energy consumption.

References

[1] Didenko N, Rudenko D, Skripnuk D 2015 Environmental security issues in the Russian Arctic *Conf. Proc. of Int. Mult. Sc. GeoConf. Surveying Geology and Mining Ecology Management, SGEM* **3** (5) 267-274
[2] Edward B 2009 A *Global Green New Deal, UNEP-DTIE* p 42
[3] Lipina SA et al 2016 *Green economy. Global development* (Moscow: Publishing House Prospekt) p 234
[4] Fedorov S G 2004 *Mining* **4** 34
[5] Metallogeny and prediction of Apatite deposits 1991.(Moscow: Nedra) p 256
[6] Nadymova M A 2016 *North and arctic in the new paradigm of world development. Lussian readings* p 674
[7] PhosAgro Publishes 2017 Integrated Report p 199
[8] Abdel-Zaher M Abouzeid 2008 *Int. J. Miner. Process.* **85** 59
[9] International Energy Agency (IEA) 2016 *Special Report Energy and Air Pollution* (Paris, France) p 266
[10] Antti Grönroos 2010 *Ultrasonically Enhanced Disintegration. Polymers* (Sludge, and Contaminated Soil. Espoo: Helsinki) p 142

Islamic diffusion of Nordic countries: Sweden

K Yu Eidemiller, [1] E A Samylovskaya[2] and R-E A Kudryavtseva

St. Petersburg State University, St. Petersburg, Russia [1]
Peter the Great St. Petersburg Polytechnic Universityб St. Petersburg, Russia [2]

katerina-samylovskaya88@yandex.ru

Abstract. The purpose of this paper is to study the Muslim migration flow to the Northern Europe in the last quarter of the 20th - first quarter of the 21st century, the pattern analysis and the genesis of the new geopolitical reality in the region, illustrated with the example of Sweden. The authors used a considerable amount of sources and literature, devoted to the study of the topic under analyses in order to accomplish the set tasks. Throughout the article, the authors relied on the advanced achievements of scientific thought, outlined in the fundamental studies of historians, sociologists, anthropologists, economists, philosophers and geographers (based on the innovations diffusion theory by a Swedish geographer T. Hagerstrand). Sweden demonstrates the paradox of the whole situation, and at the same time the main conclusion drawn from the study carried out in this paper: in 1975, the Muslim population of the Kingdom was 0.25%, in 2017, according to official data, every 10th citizen was Muslim. In Sweden only, Muslim population which equals to the population of five Iceland's populations is residing. If this tendency goes on (the events of 2015 unrepeated), then by 2025-2027, Muslim will be every 5th, and by 2035-2040, every 4th person. In 2018, according to unofficial data, the Muslim population of Sweden is balancing between 14,5 and 17,5%. 75% of all Muslims in Sweden live in two agglomerations, of which 53-55% in the metropolis of Stockholm and Uppsala, 20-25% in the capital of Denmark Copenhagen and the Swedish city of Malmö connected by the Öresund Bridge, in the absence of the actual border between the states. This circumstance creates a single region, where, according to the estimates, from 280,000 to 350,000 Muslims are living. Islamic diffusion and ghettonization which are objectively characteristic of northern territories due to their geographic, climatic, and sometimes geomorphologic features only aggravate the emerging situation.

1. Introduction

By 2025, the general economic, political and social situation in Northern Europe will consist of a mosaic of contradictions and "the unknown" prospects/ This is due primarily to such relevant problems of present times as the growth of uncontrolled migration, and as a response to this increase - the popularity of populist, rightist and extremist parties and trends in societies of the Northern Europe. The worsening of economic indicators due to the economic and financial crisis in a number of European economies, connected with the EU's sanctions policy, imposed on Russia, as a result of which Finland, Norway and Sweden are compelled to suffer losses, close enterprises and cut jobs, etc. - they all make us describe completely new realities for the Nordic countries. Other topics are the debates and referendums about the prospective entry of Finland and Sweden into NATO, and what can follow, both in terms of actions and a number of consequences - namely, the reaction of the Russian Federation to this event. In general, all the above mentioned factors tell us that the "golden age" of post-war Northern Europe is coming to an end.

Since the beginning of 2014, after a series of revolutionary events called "Arab spring", the Nordic countries have been overwhelmed by the weakly controlled waves of Islamic migration (which also took place even earlier, only on a more modest scale) from the territories of North and East Africa, as well as from the countries of South, South-West and Central Asia. The states of Northern Europe, despite their geographical remoteness, could not but share the burden of the consequences of civil conflicts in the above mentioned world regions, which from 2012 to 2013 acquired the intimidating nature of interreligious and inter-confessional wars. According to official statistics of Sweden, only in 2016, it officially registered and issued 162,877 residence permits on its territory, 4/5 of which have been in the above-mentioned regions with predominantly Muslim populations. According to official data, from 2015 to 2017, the Muslim population of Sweden, in comparison with the period from 2009 to 2011 increased by two times, unofficially by 4 times and now it amounts to no less than 17% of the total population of the country. The entire Muslim population in the Nordic countries for the same time period multiplied by 7. The available statistical data makes it possible to conclude that the Muslim part of the European population is becoming the second largest religious community after the Christian one (Catholic, Protestant or Orthodox), and Islam, respectively, becomes the second powerful religion practiced by the largest number of believers by, both in Sweden and throughout the Northern region.

The aim of the paper is to study the Muslim migration flow to the territory of Northern Europe in the last quarter of 20th - the first quarter of the 21st centuries, the analysis of the patterns and genesis of the new geopolitical reality being formed in the region, using the example of Sweden.

It is necessary to separately outline and highlight the absolute misunderstanding of the problem of Muslim communities and Islamic society development by the governments and state bodies of all the countries of Northern Europe, without exception. These are symptoms, or, to be precise, precursors of local problems, which in the future can become common for Northern Europe. Their transformation will inevitably affect cross-border cooperation and its structure after these very changes started [1]. There is a great risk that these transformations as well as their tendencies will be quite radical, as even at the initial stage they do not have a platform for compromise or manoeuvre. Two of the five Nordic countries refuse to recognize [Denmark] and accept [Iceland] the rights of Islamic communities, another [Sweden] - does not quite understand what to do and how to work with them. The other two [Norway and Finland], in fact, in the sequence of their policies and actions don’t differ much from other countries, but at least they have the resources to develop alternative, more positive scenarios for the development of Muslim communities within their own territorial and legal borders.

In order to solve the set tasks, the authors used a considerable amount of sources and literature, devoted to the study of the topic of the present work While working with the paper, the researchers relied on the advanced achievements of scientific thought, outlined in the fundamental studies of domestic and foreign internationalists, geographers, historians, sociologists, anthropologists, economists and philosophers. Political science bases of political realism, historical and historiography reviews of the Northern Europe region, relevant and necessary, reflecting the fullness of the historical perspective, civilization theories covering the problems of the migration passionaries’ movement, the influence of sectored, territorial (including migration) factors on regional labor markets, the innovations diffusion theory by a Swedish geographer T. Hagerstrand, as well as theoretical approaches to the study of transformation processes in social and humanitarian geography were additionally used. The works and scholarly writings on the directions of the theoretical foundations of multiculturalism, as well as works related to Muslim religious communities and societies, and their future in Europe and world are also taken into account.

2. Analytics

Islamic factor as a subject of domestic European politics of the neo-liberal era and the dominant ideas of the victorious multiculturalism at the turn of the 20th and 21st centuries (90s of the XX-10s of the XXI century) gave rise to the present situation, from which there are no foreseeable simple and unambiguous way outs in the present or in the future. The factor of Muslim communities in the region of Western and Northern Europe has outgrown from the category of the problem into a very real factor of life and existence of a pan-European society. In 2017, the Great (Juma) mosque conducts Friday (conciliar) services in almost every European capital, and if the capitals are metropolises, such as London or Berlin, then there are several mosques of this rank. The construction projects of the Great Mosques in Northern Europe and the Baltic region are currently widely discussed in Tallinn, Riga, Helsinki and Reykjavik — in the countries where the Muslim population is calculated in percentages, tenths, or even hundredths of percentage. In the countries of Western and Northern Europe, Muslims account for 5 to 25 % of the total population of all the capitals of the above-mentioned regions. This is a common problem for almost all the countries in the region, but each country, in every situation, up to the municipal entity, has its own peculiarity. This is the main and the biggest problem. In the states of Western and Northern Europe, there is no universal example that could serve as a template for solving a particular problem, since each case is unique within one country, and sometimes what is suitable for Stockholm no longer suits Gothenburg. Today, in all the states of Northern Europe without exception, Islam is becoming the fastest growing religion, and the Muslim community is the most dynamically developing. How it became possible, what we have now and what will possibly happen in the future – these are the problems of this thesis.

The rapid social changes that occurred in the twentieth century, together with a new round of technological progress, changed the world after the end of the Second World War beyond recognition. Thorsten Hagerstrand, a Swedish geographer who is known for his works on migration, cultural diffusion and time geography also noticed that [2]. What you could only dream of at the beginning of the twentieth century has become a reality already by its middle. The world became global, and the restrictions that still remained in it increasingly acquired formal content with time, and finally were eliminated or simply erased as the rudiments of the past eras. Colonial politics, authoritarian and totalitarian forms of government in the former metropolitan areas of Europe are replaced by a new system of values, political and socio-cultural coordinates, primarily justified by the freedom of the individual and the role of the individual in the society, regardless his characteristics, given by the nature, and peculiarities, racial or religious affiliation - humanity has acquired its oecumene [3]. Something the great philosophers of the Renaissance and Enlightenment dreamed of is finally beginning to take the final ideological shape and materialize.

However, the very theory of multiculturalism did not justify the hopes placed in it. By the end of the first decade of the 21st century, mainly within European states, a number of prominent foreign and Russian political figures believe that the theoretical basis of the hypothesis of multiculturalism seriously disconcerted with the existing reality. European politicians themselves, the leaders of European states began to state that it entered its crisis stage. This was most evident after analyzing the first results of the "Arab spring", its political and sociocultural consequences for the European communities. Scientists and researchers, politicians and experts of predominantly humanitarian specialties diverge only in assessing the degree of criticality of the theory and the European model of the political theory of multiculturalism, whether it is a crisis or rather a failure?

From 2010 to 2014, the model of the European theory of the politics of multiculturalism was met with criticism from almost all the leaders of European states sounded which described it as frankly not justifying their hopes, if less correctly, as a failure. Particularly strict experts emphasized its inefficiency and unjustified cost. The programs brought about by the policy of multiculturalism, which were the burden of social and economic responsibility on local budgets, eventually became a financial resource for the taxpayers. Speakers separately highlighted the crisis of adaptation models aimed at integrating migrants, which in fact do not assist migrants get accustomed to life in new realities, but make them become a class of new lumpens and sansculottes, living, and most often parasitizing on social benefits from the state. These models were based on the European theory of multiculturalism in completely different states of Western and Northern Europe. Therefore, based on the example and experience of their own countries, they have proved and continue to confirm its inefficiency.

In 2014, the authoritative work of the director of the Cultural Change Institute School of Law and Diplomacy at Tufts University, Professor L. Harrison, "Jews, Confucians and Protestants: the cultural capital and the end of multiculturalism" was published in the Russian translation. The main thesis of it was that the end of the period of multiculturalism is a fait accompli, and that the era of cultural racism is coming or has already come [4]. In 1991, in his work, a Swedish researcher Ake Sander already predicted the problems that his country was due to face only in 2016 [5].

Based on the above, it can be stated that the theory as well as practice, in implementation of the European multiculturalism policy is in deep crisis, if not failed. In the European society, a few years before the European migration crisis, the expression "dead brides of multiculturalism" appeared. This means a series of European girls, representatives of the autochthonous population, being raped and cruelly murdered by migrants mainly from the Muslim world. A classic and canonical example of this trend was the terrible story of a Swedish girl Elin Krantz, who, ironically, spent her whole life calling for multiculturalism in the social networks, and who even starred in the movie, where it was advocated [6]. However, it is necessary to note the fact that freedom is a relative and most often unattainable concept for girls from Muslim migrant diasporas in European societies. Their deviation from tradition, principles and practices adopted in the family and closed domestic society is often punished with death, which is called "Murder for [the sake of] honor." [7]

With that modest number of Muslims, who currently reside in Europe — continental Europe is a home for only 2.7% of all Muslims in the world — according to quite rough and generalized estimates equals 44 million. And this accounts only for 6% of the total population of Europe. We see trends, problems and vectors of their direction that are already impossible not to question, let alone to deny, which will be discussed in more details below. Year by year, the growth of the Muslim population is recorded in Europe. If we look at this problem on a global scale, then despite the wars that tear up the Islamic world - an increase in the Muslim population in the world is a global trend. In the meantime, the same wars are forcing the Muslim population to spread all over the world and go beyond natural, cultural, historical, semantic borders of the Islamic world. At present, from 38 to 41% of the world's Muslim population live in states where Islam is not the dominant religion, meaning outside the traditional Islamic world, thereby actually expanding its boundaries.

What is now happening in Northern Europe with Muslim population, is difficult to define otherwise than a conscious or unconscious social experiment that has global significance and far-reaching consequences. What is this experiment and what may be its meaning - we do not know yet. So far, from us, as from the general public, its real consequences also remain hidden. But if we only imagine that it actually exists, then at this moment it is necessary to stop, and to clarify its uniqueness and differences from all the other ones that are taking place at present.

The process of transformation of the sociocultural and political space of the Nordic countries has a number of peculiarities and stages of migration processes and is worthy of separate research and consideration. Here we turn directly to a specific and very illustrative example of Sweden. According to official Swedish statistics, only in 2016 it officially registered and accepted 162 877 immigrants on its territory, 4/5 of which come from the countries of the Muslim world. For comparison - according to the official Swedish statistics, as of July 31, 2017, the total population of Sweden was 10 065 389 people. [8] From 2011 to 2014, the Muslim population of Sweden multiplied by only 2%, and by 2015 it was at least 7.5% of the total population of the country, but as early as 2017, only according to the most conservative estimates, the Muslim population of Sweden has reached 14-16%, that is from 1,4 to 1,7 million people, which means that it has increased by two times in Sweden from 2015 to 2017.The available data makes it possible to conclude that the Muslim population becomes the second largest population in Europe after the Christian (Catholic, Protestant or Orthodox), and the fastest growing religious group on the whole European continent, and Islam, respectively, the second most faithful religion, in Sweden, as well as throughout the whole region of Northern Europe [9], with the exception of Iceland.

The number of Muslims in Sweden is only slightly inferior to their number in France, the most populous Muslim country in the EU (with the exception of the Balkan Peninsula, where Muslims constitute a significant autochthonous part of the entire population of the country). However, Muslims in large numbers began to arrive in France after the Algerian War (November 1, 1954 - March 19, 1962), and this process took almost 70 years to reach the figure that is relevant now. And it should also be noted that the Muslim population that arrived in France for almost seven decades had an almost homogeneous composition, and what is very important in this matter is their common cultural and semantic basis. However Sweden accepted approximately the same number of people in just 15 years. What is also important, a little less than half of the entire Muslim population currently living in the country - came in the last 3 years [since 2015], and they all represent completely different micro and macro worlds of the Islamic world from Western Sahara to Mindanao, and from the Volga region of Russia to Sri Lanka (Ceylon).

The difference in culture and way of life sometimes leads to much more serious contradictions. The difference in *adat - the traditional laws [pre-Islamic law]* that already brings a different "background" of problems, and sometimes these problems have such an unsolvable nature that each of the Muslim communities is forced to segregate from the dominant or trying to dominate the community or class that creates direct or indirect conflicts. As an example, one can mention that [the case of Sweden] it is undesirable for Somalis to appear in Turkish prayer houses, and those who come from the North Caucasus sometimes even live quite an isolated religious life, hidden from the eyes of strangers, which per se contributes neither to their integration, nor to their adaptation in the host societies. [10]

At the same time, the most open and integrated local Muslim religious communities, such as Tatars, Bosnians, some Turks, actually do not even represent 5% of the entire Muslim community in the region in the Nordic countries. For example, from 1917 to 1998, i.e. almost 80 years – the Tatars and Turkic-speaking ethnic groups constituted the absolute majority of the Muslim community of Finland. Eben now this fact creates the long-term problems faced by these communities in the states of the Northern Europe, and soon this problem can become common for the whole nation. If at present the so-called "old Muslims" or "old Muslim communities" is an example of the so-called "Europeanized Muslims" who currently occupy their high place of honor in the Muslim spiritual hierarchy of the Northern European societies (in fact, the place of a mediator between the ummah and the state), possibly in 5-7 years (by 2025-2027), the longest (10-15 years till 2035-2040), they will have to give their place to nationals of Muslim communities from Iraq, Afghanistan, Syria, Somalia and Pakistan. To those more significant in their number, maybe even, more weighty from the point of view of Islamic knowledge, but definitely less integrated into European society. There is little likelihood – we should underline that it is quite little –that they [the "old communities"] will be able to save their seats due to the compromise of their figures multiplied by the high level of their personal and individual integration, however, over time, the dynamics of the Muslim community growth in each of the Nordic countries certainly will not develop in their favor. And this process will be irreversible, which isn't even going to be a problem for these communities specifically and their representatives, yet another challenge for the governments of the Scandinavian states.

Norway and Sweden by 2020 will cease to be considered fully Christian countries, since the share of the practicing Christian population in them will fall below the level of 50% of the total population. Christianity in this respect becomes a ritual-ceremonial, in general and cumulative understanding - the traditional ritual part of the life of the Northern European society, nothing more. If the demographic dynamics of replacement is preserved (and in this example – particularly, replacement), the ethnic composition of the Nordic countries, to the extent that we have observed in the last 5 years, then by 2025 and after 2027 the Christian values will also be compromised in favor of more progressive ones, which correspond to the new reality and fit into a new reality for the Nordic community, but it is unlikely to be the traditional paradigm of thinking for its Muslim part. So far this has been only manifested in a few incidents like a scandal with a Christmas tree in Danish Kokkedal [11], but this is only "for now" and the key word here is "only". One way or another, in not so distant future, this will affect all spheres of life of the North European society.And we cannot but mention that the states of Northern Europe have no understanding of these events.

As on January 1, 2017, every 10th inhabitant of Northern Europe is a Muslim. In Sweden only there is a Muslim population which equals the population of five Iceland territorie. If the same trend continues (without repeating the events of 2015), then by 2025-2027, it will be every 5th, and by 2035-2040. each 4th (see Table 4).

Table 1. The number of Christians in the Nordic countries in 2006 and as of January 1, 2017 (in%) *

Country	Number of Christians as of January 1, 2017			Number of Christians in 2006
	Lutherans	Catholics	Other Christian Believers	
Kingdom of Denmark	75,9% [12]	1,0%	-	82,6%
Republic of Finland	71% [13]	0,2%	2,5%	84,2%
Republic of Iceland	69,8% [14]	2,5%	11,3%	76%
Kingdom of Norway	71,5/53% [15]	1,6%	2,7%	82,7%
Kingdom of Sweden	41,0% [16]	2,0%	9,0-10,0%	84%

(Compiled by the authors after processing statistical evaluation materials)

Table 2. The number of Muslims in the Nordic countries in 1975, 1995, 2006, as of January 1, 2016 (in%)

Country	Number of Muslims in the Nordic countries			
	1975	1995	2006	2016
Kingdom of Denmark	0,6%	2,1%	4,2%	5%/9,4-11,2%*
Republic of Finland	0,001%	0,2%	1%	3,5-4,7%
Republic of Iceland	-	-	0,01%	0,33-0,67%
Kingdom of Norway	0,001%	1,3-1,5%	3%	5,7-7,3% / 25,2%**
Kingdom of Sweden	0,25%	2,3%	5%	10,7-14,5%

(Compiled by the authors after processing statistical evaluation materials)

** The Danish authorities do not register the religious affiliation of people other than members of the National Church (the Church of the Danish People), while local researchers take into account only members of registered communities.*

*** 25.2% of all people practicing in Norway.* [17]

(Compiled by the authors after processing materials of national statistical evaluation bureaus)

Table 3. Religions in the Nordic countries as of November 1, 2016 (in%) according to official statistics

Country	Christianity			Islam	Buddhism	Judaism	Atheists or agnostics
	Lutherans	Catholics	Other Christian Believers				
Kingdom of Denmark	76,9	1	-	4,8 - 5	-	-	13,0
Republic of Finland	72,9	0,2	2,5	3,5-4,7	-	-	24,3
Republic of Iceland	71,5	3,6	7,3	0,33-0,67	0,3	0,1	5,7
Kingdom of Norway	72,9	2,5-5	2,7	5,7-7,3	0,2	0,01	13,6
Kingdom of Sweden	63,2	2	9,0-10,0	10,7-14,5	1	-	17/46*

(Compiled by the author after processing materials of national statistical evaluation bureaus)

**Phil Zuckerman, Associate Professor of the Department of Sociology, Pittser College, writes that in recent years, several scientific and academic sources have posted the data that the percentage of atheists in Sweden is between 46 and 85 %, although officially, only 17% of respondents identified themselves as "atheists".* [18] *Something similar, which looks like a trend, with figures that vary in percentage only, can be clearly seen in Iceland and Norway.* [19]

Table 4. The share of Muslims in the Nordic countries from 1975 to 2050, in% of the total population

Country	Number of Muslims in the Nordic countries				
	1975	**1995**	**2006**	**2016**	**2050**
Kingdom of Denmark	0,6%	2,1%	4,2%	5% /9,4 -11,2%	18 - 25%
Republic of Finland	0,001%	0,2%	1%	3,5 - 4,7%	7 - 9%
Republic of Iceland	-	-	0,01%	0,33 - 0,67%	1 – 1,5%
Kingdom of Norway	0,001%	1,3-1,5%	3%	5,7 - 7,3%	13-15%
Kingdom of Sweden	0,25%	2,3%	5%	10,7 - 14,5%	23 - 27%

(Compiled by the authors after processing materials of national statistical evaluation bureaus)

3. Share and concentration of the Muslim population in Sweden, in dynamics (1997-2030)

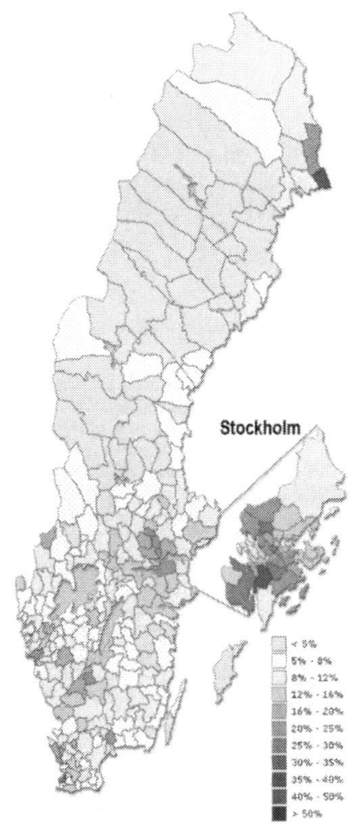

Figure 1. Share and concentration of the Muslim population in Sweden in 1997-2000.

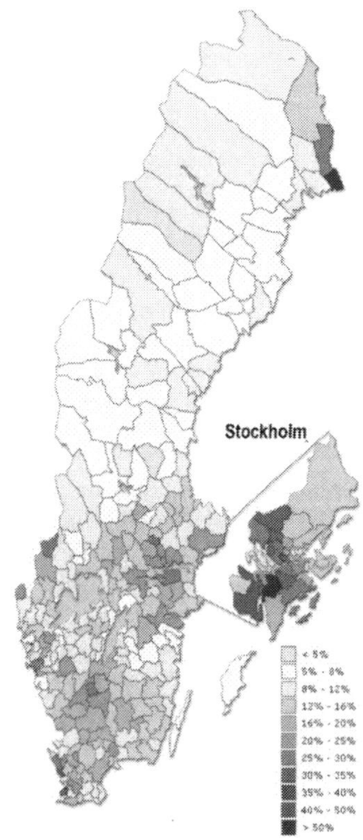

Figure 2. Share and concentration of the Muslim population in Sweden in 2010

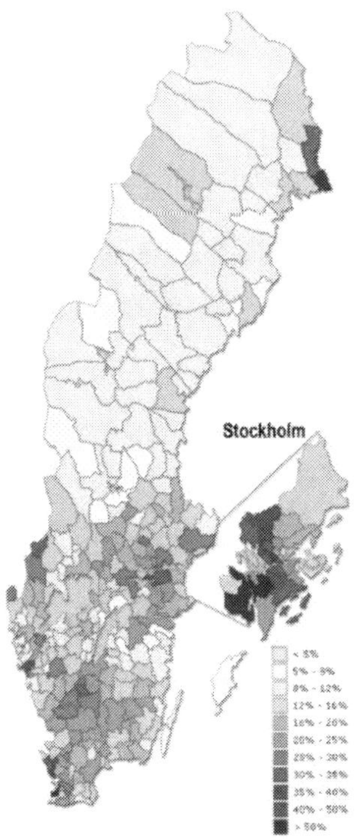

Figure 3. Share and concentration of the Muslim population in Sweden by 2020

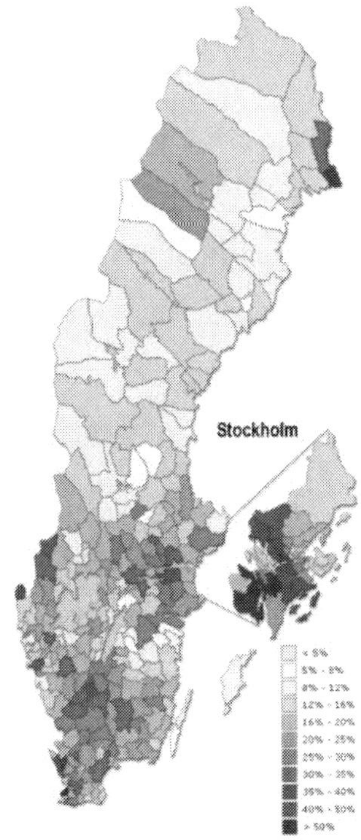

Figure 4. Share and concentration of the Muslim population in Sweden by 2030

4. Conclusion

Majorconclusions of this paper:

- , The problem of spontaneous growth of the Muslim population, communities and organizations can be felt in Northern Europe most acutely, problematically and symptomatically — like nowhere else. This is the only region of the world in which Islam and Muslim communities witnessed a significant, serious and confident growth in the XXI century;
- As of January 1, 2017, every 10th resident of the Northern Europe is a Muslim. Only Sweden hosts Muslim population comparable to the population of five Iceland territorie. If the same trend is preserved (not repeating the events of 2015), then by 2025-2027 a Muslim will be every 5th, and by 2035-2040. every 4th citizen;
- Looking at Sweden, it is possible to imagine the paradox of the whole situation: in 1975 the Muslim population of the Kingdom was 0.25%; in 2017, according to official data, it is every 10th citizen, according to unofficial data, the Muslim population of Sweden balances between 14,5 and 17,5%. 75% of all Muslims in Sweden live in two agglomerations, of which: 53-55% in the metropolis of Stockholm and Uppsala, 20-25% in the one that forms the capital of Denmark - Copenhagen and the Swedish city of Malmö, connected by the Öresund Bridge in the absence of the actual border between the states. This circumstance creates a single region in which, according to estimates, from 280,000 to 350,000 Muslims live. Islamic diffusion and ghettonization of the territory - objectively characteristic of northern territories due to their geographic, climatic, and sometimes geomorphological features — only makes the emerging situation worse;

- The conflict potential of the whole so-called "Islamic world" and its rapid internal transformation guarantee a stably high level of migration to Northern Europe with the prospect of these flows' increase;
- Deconstruction and reformatting of the so-called "Islamic world" from the outside, including the "soft power", is likely to turn into a complete catastrophe for it in the paradigm of current realities with all the ensuing consequences, both for the whole world and the Nordic region in particular;
- In most countries of Western and Northern Europe, local Muslim "parallel societies" already exist or are being formed. Theirinfluence on the state level will increase with the time while maintaining the present dynamics.

References

[1] Eydemiller K Y 2016 Transformation of the Confessional Structure of the Nordic Countries as a Destabilization Factor in the Arctic Region *Natural and Cultural Heritage of the Arctic. Coll. of Articles on the Materials of the Scientific and Practical Conf. November 10-13, 2016* 106-116

[2] Hannerberg D and Hagerstrand T 1957 Migration in Sweden: Symposium. *Lund Studies in Geography: Series B, Human Geography* **(13)**

[3] Gladkii Y N, Grigoriev Al A and Yagya V S 1990 The Horizons of the Oecumene. *Mind Studies the World* 271

[4] Harrison L 2014 *Jews Confucians and Protestants: Cultural Capital and the End of Multiculturalism* (Moscow: Thought) p 286

[5] Sander A 1991 *The Road from Musalla to Mosque. Reflections on the Process of Integration and Institutionnalisation of Islam in Sweden* (Gothenburg: Department of Philosophy, University of Gothenburg)

[6] Krantz E "We Like Diversity"! Available from: http://www.dailymotion.com/video/x2wd2md [Accessed 23th January 2018]

[7] RT in Russian: Murder of Honor: 5, 000 Victims Annually Worldwide. Available from: https://russian.rt.com/article/313292-ubiistvo-chesti-5-tys-zhertv-ezhegodno-po [Accessed 23th January 2018]

[8] Population statistics Available from: URL:http://www.scb.se/en/finding-statistics/statistics-by-subject-area/population/population-composition/population-statistics/ [Accessed 23th January 2018]

[9] Göran L 2017 Sweden *Yearbook of Muslims in Europe* **9** 645-658

[10] Bangstad S 2013 Eurabia comes to Norway *Islam and Christian-Muslim Relations* **3** 369-339

[11] Ermolaeva N 2012 Europe: Abduction of the Christmas Tree URL: Available from: https://rg.ru/2012/12/25/yolka-poln.html [Accessed 23th January 2018]

[12] Folkekirkens medlemstal *Kirkeministeriet* Available from: http://www.km.dk/folkekirken/kirkestatistik/folkekirkens-medlemstal/ (Accessed: 23.01.2018)

[13] Kirkkoon kuuluu 71 prosenttia vaestosta *Suomen Evankelis-Luterilainen Kirkko* Available from: https://evl.fi/uutishuone/tiedotteet/-/items/item/17504/Kirkkoon+kuuluu+71+prosenttia+vaestosta [Accessed 23th January 2018]

[14] Populations by religious and life stance organizations 1998-2017 *Statistics Iceland* Available from: http://px.hagstofa.is/pxen/pxweb/en/Samfelag/Samfelag__menning__5_trufelog/MAN10001.px/table/tableViewLayout1/?rxid=2c4e382d-5cb3-4b12-b9a0-9b482ecdbe01 [Accessed 23th January 2018]

[15] Church of Norway *Statistics Norway* Available from: http://www.ssb.no/en/kultur-og-fritid/statistikker/kirke_kostra/aar [Accessed 23th January 2018]

[16] Svenska kyrkan i siffror Available from: https://www.svenskakyrkan.se/statistik [Accessed 23th January 2018]

[17] Members of Religious and Life Stance Communities outside the Church of Norway, by Religion Life Stance. Per 1 January *Statistics Norway* Available from: https://www.ssb.no/en/kultur-og-fritid/statistikker/trosamf/aar/2015-11-25?fane=tabell&sort=nummer&tabell=245519 [Accessed 23th

IOP Conf. Series: Earth and Environmental Science **180** (2018) 012005 doi:10.1088/1755-1315/180/1/012005

January 2018]
[18] Greenfield N 2009 The Virtues of Godlessness *SSRC* Available from: http://blogs.ssrc.org/tif/2009/02/02/the-virtues-of-godlessness/ [Accessed 23th January 2018]
[19] Steinfels P 2009 Scandinavian Nonbelievers, Which Is Not to Say Atheists *The New York Times* Available from: http://www.nytimes.com/2009/02/28/us/28beliefs.html?pagewanted=all [Accessed 23th January 2018]

Justification of project and operation modes of hybrid energy complexes for arctic conditions

V V Elistratov, M V Diuldin and R S Denisov

Peter the Great St. Petersburg Polytechnic University, St. Petersburg, Russia

denisov.roman.90@gmail.com

Abstract. Currently, one of the priorities based on the "Energy strategy of Russia's development to 2030" is the development of energy infrastructure of the Arctic and Far Eastern regions of Russia, which geographically make up about 65% of the country and located in the areas of decentralized power supply. The total installed capacity of diesel power plants (DPP) operating in the northern regions are more than 500 thousand kW with the production of about 2.5 billion kW·h, which requires the consumption of about 1 million tons of diesel fuel per year. Diesel fuel delivery to these regions carried out in the framework of the "northern" delivery. The article considered methods of hybrid systems based on renewable sources design, allowing optimize technical and economical parameters and increase the penetration of diesel fuel imported to the northern territories to provide consumers with the energy. The first method is a 3-level sequence of the wind energy resources (WER) assessment at a specific point of the energy complex location, for enhancing the reliability of which reanalysis database of satellite observation are used and most accurately describes the wind flow in the Russian northern regions. The second method is the operation modes justification of the power distribution between diesel generators at the diesel power plant. Both methods have been tested at the existing hybrid system 1000 kW (800 kW – diesel part, 200 kW – wind part) in the village Amderma (Nenets Autonomous District, Russia).

1. Introduction

Almost the entire Russian Arctic territory is located in the isolated (autonomous) power supply zone and mainly provided with electricity from diesel power plants (DPP) operating on imported fuel.

The cost of energy production for such DPP is 0.25-2.0 Euro / kWh, which is much more expensive than in the areas of centralized power supply. At the same time, the wind energy potential in autonomous power supply areas is high and can be used by creating autonomous hybrid systems - wind-diesel power plants (WDPP).

Parameters and operating modes justification of such WDPP is an important scientific and practical task, the competent solution of which will ensure efficient and modern power supply to consumers, a high penetration of diesel fuel, a reduction of diesel delivery volume from the north, saving of a diesel engine life and reducing the economically justified tariff. The solutions of these tasks consist of two main subtasks: 1 – reliable assessment of the available wind energy resources in limited natural and climatic information conditions; 2 - parameters and operating modes of the hybrid systems justification to ensure the maximum replacement of diesel fuel.

2. Description of the approach for estimating wind energy resources using reanalysis databases for the Arctic territories

The assessment of wind energy resources in Russia is based on methods [1-6], using materials of regular observations from the meteorological stations of the USSR State Committee of Hydrometeorology and the Russian Hydrometeorological Comittee. Available datasets from meteostations are often insufficient due to the low density of the meteorological network. However, due to the low density of the meteorological station network, the available data are often insufficient, since the distance between them can significantly exceed the characteristic scale of the spatial variability of the wind characteristics. The average coating density of hydrometeorological stations (HMS) in Russia is about 51582 km^2, which corresponds to the distance between the network HMS stations about 227 km [7]. In the Far North the meteorological coverage density is even lower and the value of wind resources at the location of the wind farm remote from the HMS may significantly differ from the values measured on it.

In the lack of high-quality meteorological information the authors proposed a methodology of a three-level wind potential assessment [4] to improve the accuracy of electricity generation forecast at the wind farms, using the second-level mesoscale evaluation of the reanalysis databases developed by various international organizations and institutions. Reanalysis data is based on satellite observations processed in global numerical weather prediction models (NWP) and represent as datasets of synthesized values (wind speed and direction) at each point of mesh. The main disadvantage of using reanalysis data is different validation in different regions of the world [8, 9].

From the existing reanalysis databases were selected covering all territory of Russia (CFSR [10], MERRA [11], NCEP/NCAR [12]) and data of the secondary reanalysis (Met Office [13]) and Vortex [14]), which scales traditional reanalysis data with using NWP models with higher resolution.

The algorithm for resources estimating using reanalysis bases is shown in Figure 1.

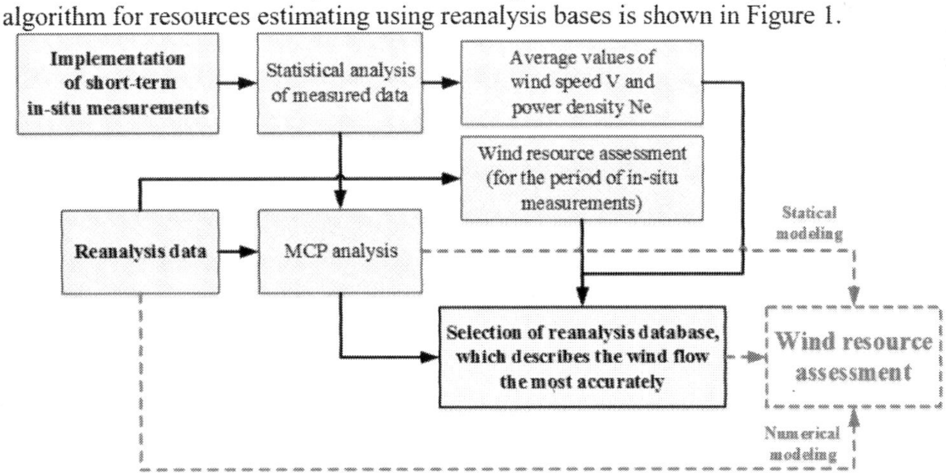

Figure 1. The algorithm of wind resource assessment based on reanalysis data.

Algorithm is conducted in several stages:

2.1. Implementation of short-term in-situ measurement

Observations were carried out with specialized wind monitoring complex in accordance with international standards [15, 16].

2.2. Statistical analysis of measured data. Calculation average values of wind speed V and power density Ne during the measurement period

During statistical processing measured data was filtered by invalid values in cases of anemometer icing and tower shading. The power density of wind flow in each moment is determined by the equation:

$$Ne_i = \frac{1}{2} \cdot \rho \cdot V_i^3, \quad (1)$$

where ρ - air density, V_i – wind speed in each moment.

2.3. Conducting Measure-Correlate-Predict (MCP) analysis with a processed series of full-scale measurements, determining correlation coefficients and synthesizing the long-term wind statistics.

MCP analysis allows to estimate a correlation of time series by linear regression. Correlation degree is determined by the coefficient r^2, characterizing the dispersion (standard deviation) of the measured values from the line that describes the relationship between the time series [17]. Measurements in compared datasets (target and reference time series) may be done at different heights and in different points of terrain. The main conditions for the selection pair of time series is their homogeneity and location within one climate zone.

A comparison of the series using the MCP analysis can be performed with different intervals, which significantly affect the resulting value of the determination coefficient. In case of very small comparing intervals the ranks may differ due to various micro-climatic factors in the measuring point. In case of large comparing intervals, the characteristics of wind speed stroke can be flatten and not reflect the real picture. The optimal value of comparing interval is determined by the empirical relationship [18]:

$$\Delta t \approx \frac{D}{v}, \quad (2)$$

where D is distance between points of compared datasets [m], ν is mean wind speed [m/s].

Interval comparison is assumed to be optimal, except the cases with greater frequency measurements in the series.

The recommended value of the determination coefficient r^2 in the international practice of wind monitoring is 0.7-0.8, which ensures a good comparability of the two time series and is considered to be acceptable to synthesize a long-term wind statistics with high accuracy [19].

Because of the small and of similar value of correlation coefficients for different reanalysis database and taking into account the scope of further use of these data (direct evaluation of the WER at the point without comparison with the field observations), it is advisable to compare the WER value estimates with the period of in-situ measurements.

2.4. Annual wind resource assessment (for the period of in-situ measurements) based on reanalysis data using numerical simulations

For wind resource assessment based on reanalysis data, geographically remote from a particular point, microscale modeling is performed in the WindPRO software package, taking into account the influence on the wind flow of the relief and the roughness of the underlying surface. To enable direct comparison of results with the wind monitoring results, the measurement period was assumed equal to the period of field observations [20].

For each reanalysis database calculations were made separately for each grid node, and also based on several series simultaneously (using weight coefficients depending on the distance from the calculated point).

For the time series, geographically located at the same point (secondary reanalysis data), conducted wind speed distribution by the gradations description of Weibull distribution and its further vertical extrapolation to the respective vertical profiles.

Assessment of the gross wind potential is major step in the design of wind farm, however the large differences in wind power estimates could significantly affect the estimates of annual wind farm power output due to the non-linearity of the wind turbine power curves. Therefore, the generation of electricity for a specific wind turbine located at the point of in-situ measurement was also calculated.

As the wind turbine was considered nED 100 [21] manufactured by Norvento enerxia (Spain) with the installed capacity of 100 kW and a tower height of 29.5 m. The wind turbine type was selected out of using in the northern regions wind turbines with small capacity and an extended operating temperature range [22].

2.5. Analysis of the results obtained. Selection of reanalysis database, which describes the wind flow in the northern regions of Russia the most accurately

Choosing the most suitable reanalysis database is based on an analysis of two criteria: the determination coefficient and the estimates of wind resources obtained by numerical simulation. The determination coefficient reflects the considered series correlation with the in-situ measurement data, identifies the series most suitable for synthesizing a long-term statistics based on short-term observations (statistical modeling). The use of in-situ measurement data at a specific location allows to verify the series obtained at numerical modeling and thereby increase the reliability of the WER estimation. The introduction of the correction factors for V_{ave} and Ne allows to evaluate the WER in the proposed construction site directly on the basis of the reanalysis data (numerical modeling) without in-situ measurements.

2.6. Wind resource assessment based on long-term wind statistics with using statistical and numerical modeling. A comparison of the values with the existing estimates (based on data from the meteorological station)

Statistical modeling is synthesizing the long-term wind statistics using the MCP analysis based on short-term in-situ measurements (target data) and reanalysis data time series with the highest correlation. In this case synthesized long-term wind statistics presents the characteristics of wind flow in the point and at the height of wind monitoring.

Numerical modeling is similar to paragraph 2.4 and based on long-term time series from selected reanalysis database.

3. Distribution of power between the generating equipment of diesel power plant and wind power plant

The main criterion of functioning when designing a WDPP with a high penetration in arctic conditions is the minimization of the consumption of diesel fuel while providing a balance of production and consumption of electric energy. To achieve this criterion, it is required to ensure maximum power generation for wind turbines in all operating modes.

The main characteristic in the distribution of power in the energy system is the power balance at each given time, which ensures the stabilization of the frequency and voltage of the autonomous network and is given by formula:

$$\begin{cases} \Sigma P_S = \Sigma P_L \\ \Sigma Q_S = \Sigma Q_L \end{cases} \tag{3}$$

where ΣP_S, ΣQ_S – active and reactive power of all sources, kW; ΣP_L, ΣQ_L – active and reactive load powers, kW.

If there is an excess of power +ΔP, this value is useful for charging the accumulating system (AS). For this purpose, it is checked the condition "state of the charge (SOC) = 100%". By 100%, the excess capacity can be reset to a ballast load or to reduce the output from the wind farm by the pitch control of the blades. Underestimation of output is useful when increasing the load power, as a frequency or spinning reserve.

When there is a shortage of power, amount -ΔP is created, covered by switching off the ballast load, connecting the spinning reserve or the stored energy from the AS. In view of the high inertia of the diesel-generator system, the power redistribution functions at the time of switching off the diesel generators are assigned to the AS with a bi-directional power converter. When predicting a continuous shortage of power, the diesel generators is sequentially switched on, beginning with smaller units of smaller capacity, due to the lesser inertia of the latter. When the condition "SOC is less than the minimum", a part of the power from the DPP goes to the charge of the AS.

In this moment, the optimal distribution of power between the generators makes using DPP more efficiently, additionally saving fuel. At the same time, it prevents the work of equipment in the non-recommended area.

The objective function for this task is to minimize fuel consumption Q and reduce the number of cycles of on/off DPP:

$$\begin{cases} Q \rightarrow \min \\ n \rightarrow \min \end{cases} \quad (4)$$

Since the diesel generator in the hybrid system with a high penetration at any moment of time is in the off state, the required power value denoted in the previous section ΔP is covered either by one diesel generator or by a combination of generators from condition:

$$P_{i50} < \Delta P < P_{i90} \quad (5)$$

where P_{i50}, P_{i90} – power of i-generator with loading ration between 50% and 90% (security of claim of energy reserves in case of system fluctuations without loss of the operational efficiency of the DPP).

When performing this condition it is switched on only the most efficient generator, which saves the runtime of other generators, which will reduce the number of outages in the system and increase the operating time to failure. The determination of the most effective installation is based on a comparison of the approximate functions of the fuel consumption on the unit's power Q(P).

$$Q = a \cdot P^b + c, \quad (6)$$

If condition (5) is not fulfilled, the optimal distribution must be considered as a solution of the equation system:

$$\begin{cases} Q = a_1 \cdot x_1^{b_1} + c_1 + a_2 \cdot x_2^{b_2} + c_2 + \cdots + a_n \cdot x_n^{b_n} + c_n \\ x_1 + x_2 + \cdots + x_n = \Delta P, \\ 50\% < x_i < 90\% \end{cases}, \quad (7)$$

To solve such an equation system, it is apply the Lagrange multiplier method with the Kuhn-Tucker condition [22]:

$$\begin{cases} \frac{\partial L}{\partial x_1} = a_1 \cdot b_1 \cdot x_1^{b_1 - 1} + \lambda = 0, \\ \frac{\partial L}{\partial x_2} = a_2 \cdot b_2 \cdot x_2^{b_2 - 1} + \lambda = 0, \\ \ldots \\ \frac{\partial L}{\partial \lambda} = x_1 + x_2 + \cdots + x_n - \Delta P = 0 \end{cases}, \quad (8)$$

By solving this system, we obtain a solution for the optimal distribution of active power at the diesel power plant.

However, with a continued increase in power, it is required to recalculate the Lagrange function (8), which can cause additional outages in the system. For example, the variant of the possible loading of DPP is shown in Figure 2. For some value of ΔP1, DG#2 is in the off state. If the Lagrange function is recalculated for a new value ΔP2, it may occur a situation when it is necessary to turn off the DG#3 and turn on the DG#2, which causes additional commutations in the system and increasing switching times. Thus, if the continuous operation of the DPP is occur, the system uses a "greedy" algorithm, the basis of which is to maximize the use of already included equipment (Figure 2).

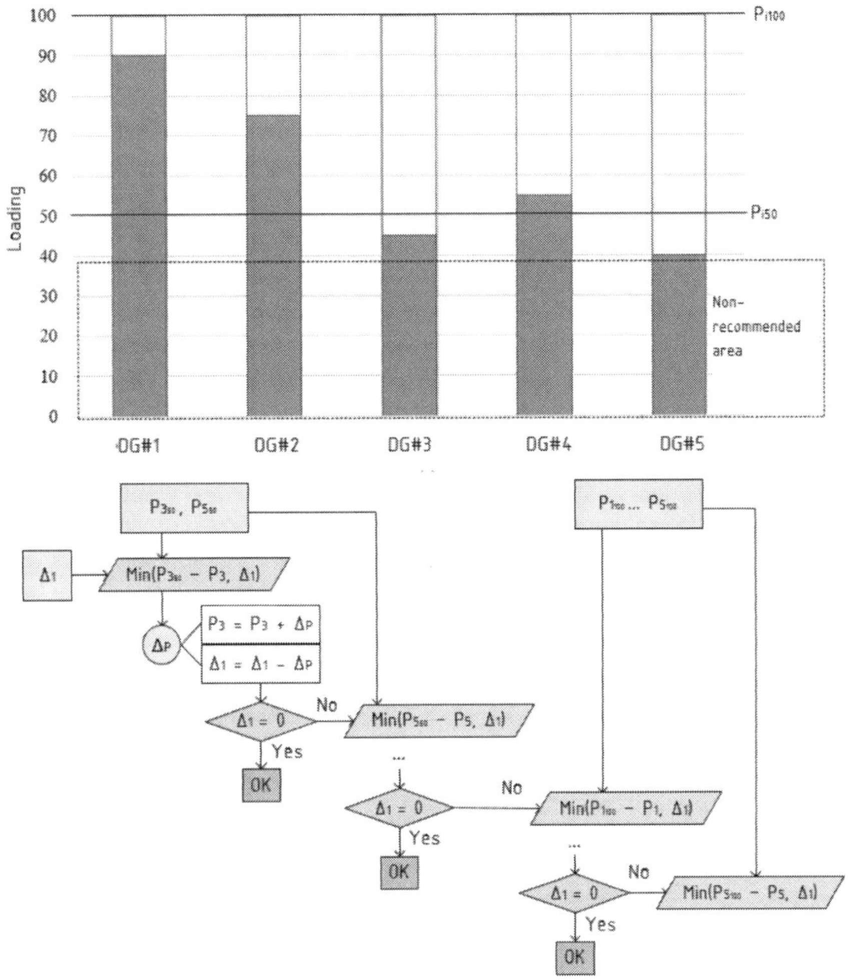

Figure 2. The algorithm of distribution of power at the DPP

When the power is increased by any amount of Δ1, the system analyzes the switched-on DGs, selecting those that operate in the non-recommended (low-efficiency) area. To ensure an increase in power by an amount Δ1, the system alternately increases the power of each such DG to 50%, choosing them in order of efficiency. Thus, the most efficient diesel fuel consumption is the first to be loaded. If this increase in power at the DPP does not cover the value of Δ1, the DGU at the current capacity is once again evaluated for efficiency. Thus, the system continues to load the most efficient diesel fuel consumption to 100%. An example of the algorithm is shown in Figure 2. Based on the example of a DES consisting of 5 DGUs, located in the figure from left to right in efficiency, an algorithm is shown for loading the plant with a power increase by Δ1.

If the power is reduced by Δ2, the system reduces the load of the most inefficient fuel consumption of the diesel generator from the end to the value of 50%, so that it remains possible to reduce the load to 40% with an additional reduction in power.

4. Research results on the WDPP in the village Amderma

Approbation of methods was carried out during the design and operation of the WDPP in the village Amderma in the Russian Arctic area [4, 19, 23]. As a result of filter observations of wind monitoring, time series of wind speed and direction at a height of 40.8 m was obtained. Average value of the wind speed is 8.23 m/s. Mean wind flow power density is 666 W/m^2. The data were used in the design of

WPP and justification of its production. The vertical profile of wind speed distribution is described by a power law and corresponds to the profile index $\alpha = 0{,}196$.

Based on the results of numerical simulations, assessment of gross wind resources based on reanalysis data significantly underestimated (13 to 64%). This exceeds the accuracy of modeling that characterizes the quality of reanalysis database as insufficient for using without a comparison with in-situ measurements or climatological data directories. The greatest uncertainty in the estimates is given by secondary database reanalysis (Vortex and MetOffice), that justified higher resolution mesoscale modeling data series. Such datasets does not recommended to be considered when microscale numerical modeling can be proceeded. Despite the fact that the best estimate of the average accuracy rate was obtained from one of the series CFSR, the simulation results based on each point of the mesh differ. Most similar values of wind resource estimates have reanalysis database MERRA, indicating a high precision used in its NWP simulations digital models of orography and roughness.

As expected, the errors of technical wind resource assessment were significantly less than errors of gross values. Apart from the Vortex series, all datasets give the error not more than 20%, it is comparable with the errors of numerical modeling. The closest indicators (8-9% error) have reanalysis databases CFSR and MERRA. The proximity of the simulation results based on a series MetOffice is attributed to random coincidences due to the intersection of the vertical profiles at the height of the wind power plant tower. The solution to this problem can be solved in two cases:

- Establishment of correction coefficients for the gross (a_g) and technical (a_t) wind resources based on relationships:

$$a_g = \frac{Ne_{vic}}{Ne_i}, \qquad a_t = \frac{W_{vic}}{W_i}, \tag{9}$$

where N_{ei}, N_{vic}, W and W_{vic} are estimates of the wind flow power density (Ne, W/m^2) and annual power output (W, MWh/year) based on reanalysis data and in-situ measurements at the same period respectively.

- Establishment of a correction coefficient for wind speed (a_v). In this case, the numerical simulation of the wind flow is based on the modified time series of wind speed in each point of mesh for corresponding reanalysis database.

$$a_v = \frac{V_{vic}}{V_i} = \sqrt[3]{\frac{Ne_{vic}}{Ne_i}} = \sqrt[3]{a_g}, \tag{10}$$

where V_i и V_{vic} are mean values of wind speed for reanalysis data and in-situ measurements at the same period respectively.

Correction coefficients were calculated for reanalysis databases CFSR and MERRA. The values of gross and technical wind resources in point of wind monitoring were calculated with using these corrections, except NeMCP and WMCP, which were defined with using time series synthesized on the base of in-situ measurements using MCP analysis (table 1).

Table 1. Correction coefficients for reanalysis databases CFSR and MERRA and wind resource estimates based on it for the village Amderma.

Reanalysis database	ag, p.u.	at, p.u.	av, p.u.	Ne(ag), W/m^2	W(at), MWh/year	Ne(av), W/m^2	W(av), MWh/year	NeMCP, W/m^2	WMCP, MWh/year
CFSR	1.52	1.08	1.15	783	442	709	462	678	442
MERRA	1.43	1.09	1.13	723	422	734	456	595	430

The consumption of the DGs at the station due to the application of the methodology decreased by an additional 10% per year [23], thus reducing the cost of own needs by three times (from 510 to 160 ths. kWh per year). Saving diesel fuel per year for WDPP without the use of accumulating systems is 303 thousand liters per year (by 40%), maintenance costs for DES disappeared 1.3 times, and CO_2 emissions decreased by 600 tons per year.

5. Conclusion

In order to improve the accuracy of the natural and technical resources assessment in the absence of meteorological observations in the proposed location of wind farms vicinity in the Nenets Autonomous District, it is recommended to use the MERRA reanalysis database and apply a correction factor for wind speed equal to 1.13. Calculation of correction factors for wind speed and gross technical potential of the other Russian Arctic regions recommended to be carried out similarly according to the procedure described above.

Reanalysis databases allows to obtain short-term (up to 10 minutes) long-term series of wind statistics, in contrast to the use of reference data, containing integral characteristics of wind speed and wind energy resources. The use of wind speed series obtained allow to justify the choice of WDPP equipment and to improve the accuracy of the electricity production forecast at the wind farm and the operation modes of its cooperative work with the DPP.

An algorithm for the distribution of power at the DPP was developed to reduce the inefficient fuel consumption. The mathematical apparatus of this algorithm is adapted to the DPP practical implementation, taking into account the reduction of the diesel sets in the operating time associated with the number of switching on/off the particular diesel generators. The developed methodology allows to further reduction of the diesel fuel cost at DPP, increasing the operating time of the equipment.

Application of proposed methods in the design of WDPP for Arctic conditions and optimization of its operating modes allows increasing the share of diesel fuel replacement, which is reflected in the amount for subsidies reduction when covering economically justified tariff for the settlements located in the Arctic regions. For the WDPP project in the village Amderma economically justified tariff is reduced by 38% (about 750 thousand $).

References

[1] Borisenko M M *et al* 1991 *Carrying out prospecting wind resource assessment for substantiation schemes of siting and design of wind turbines. Methodological guidelines* (Moscow: Ed. Roshydromet) p. 55 (in Russian)

[2] Nikolaev V G, Ganaga S V, Kudryashov Yu I 2008 *National inventory report of wind energy resources in Russia and methodological basis for their determination* (Moscow: Atmograf) p. 584 (in Russian)

[3] Starkov A N, Landberg L, Besroukikh P P, Borisenko M M 2000 *Russian Wind Atlas* (Moscow: Mozhaysk-Terra) p. 560 (in Russian)

[4] Elistratov V V 2016 *Renewable energy* (St. Petersburg.: Science) p. 424 (in Russian)

[5] Elistratov V V *et al* 2010 *Climate factors of renewable energy sources* (St. Petersburg.: Science) p. 235

[6] Elistratov V V, .Kudryasheva I.G., Pilipets P.A. 2015 *Energy efficient solutions of power supply in north regions* Applied Mechanics and Materials **725-726** 1463-1469

[7] Slivkanich M, Savchenko D, Diuldin M 2015 Diesel power plants power output penetration by using wind energy *Science and Innovation in Technical Universities: Proceedings of the 9th All-Russian Forum of Students, PhD Students and Young Scientists (*St. Petersburg) 151-153 (in Russian)

[8] Franke J, Hellsten A., Schlünzen H., Carissimo B. 2007 *Best practice guidelines for the CFD simulation of flows in the urban environment. Quality assurance and improvement of micro-scale meteorological models* (Cost Action, Belgium) p. 732

[9] Tinker J, Lowe J, Holt J, Pardaens A, Wiltshire A 2015 *Validation of an ensemble modelling system for climate projections for the northwest European shelf seas* Progress in Oceanography **138 A** 211-237

[10] Suranjana S *et al* 2012 The NCEP Climate Forecast System Reanalysis *Bulletin of the American Meteorological Society* **91** 1015–1057

[11] Rienecker *et al.* 2011 *MERRA: NASAs modern-era retrospective analysis for research and applications* J. Clim., No. 24, pp. 3624–3648

[12] Kalnay E *et al* 2013 The NCEP/NCAR 40-Year Reanalysis Project *Bulletin of the American*

Meteorological Society **77** 437-471
[13] Williams K D *et al* 2018 The Met Office Global Coupled Model 3.0 and 3.1 (GC3.0 and GC3.1) Configurations *Journal of Advances in Modeling Earth Systems* **10** (2) 357-380
[14] Byrne N J *et al* 2017 Nonstationarity in Southern Hemisphere Climate Variability Associated with the Seasonal Breakdown of the Stratospheric Polar Vortex *Journal of Climate* **30** (18) 7125-7139
[15] International Electrotechnical Commission (IEC) Standard 61400-12-1 First edition 2005-12: Wind turbines, Part 12-1: Power performance measurements of electricity producing wind turbines
[16] MEASNET Procedure: Evaluation of Site Specific. Wind Conditions. Version 1. November 2009
[17] Rogers A L *et al* 2005 Comparison of the performance of four measure-correlate-predict algorithms *Journal of Wind Engineering and Industrial Aerodynamics* **93 (3)** 243-264
[18] Brower M C 2012 *Wind Resource Assessment: A Practical Guide to Developing a Wind Project* (A John Wiley&Sons, Inc.) p. 265
[19] Elistratov V V, Diuldin M V, Slivkanich M A 2014 Measurement of Wind Flow Characteristics for Wind Resource Assessment Using Sodar *International Scientific Journal Life and Ecology* 1 21-22
[20] Petersen E, Mortensen N, Landberg L, Hojstrup J Frank H 1997 *Wind Power Meteorology* (Riso National Laboratory, Roskilde, Denmark) p. 342
[21] Elistratov V, Konishchev M, Fedorov M. 2017 Optimization of Power Supply of the Circumpolar Territories on the basis of Renewable Energy Sources *2017 Intern. Conf. on Industrial Engineering, Applications and Manufacturing, ICIEAM 2017*
[22] Elistratov V V, Denisov R S, Konishchev M A, Knezhevich M 2014 Problems of construction Wind-Diesel Power Plants in harsh climatic conditions *Journal of Applied Engineering Science* **12** (1) 29-36
[23] Elistratov V, Denisov R 2017 Justification of wind turbine operation and management modes for russian arctic conditions *2017 Intern. Conf. on Industrial Engineering, Applications and Manufacturing, ICIEAM 2017*

The environmental impactcaused by developing energy resources in the Arctic region

S Agarkov, T Motina and D Matviishin

Murmansk State Technical University, Murmansk, Russia

motinatn@mstu.edu.ru

Abstract. The Arctic region has been attracting a great deal of interest from numerous researchers. Developing the oil and gas resources of the Arctic shelf is an important task for the national economy, especially since the area supply with hydrocarbons is decreasing. At the same time, preserving the integrity of Arctic marine ecosystems is of paramount importance taking into account the vulnerability of the Arctic seas. In view of the current trends, the Arctic region is rightfully considered strategically significant for further sustainable development of the Russian Federation. Intensified industrial development of hydrocarbons in the Arctic shelf of the Russian Federation lends urgency to assessing the impact of developing energy resources on the environment. The main conditions for successful development of oil and gas resources on the Arctic shelf of Russia are economic efficiency and environmental safety. Biological diversity (of species) is one of the main indicators of ecosystem sustainability. Species diversity may be affected not only by climate change, but also by the environmental changes associated with anthropogenic impacts, such as products of hydrocarbon exploration, extraction and refinery, heavy metals and other metallurgical wastes. Using approved environmental monitoring methods the paper presents the results of the assessment of the current environmental status and biodiversity in industrial development areas illustrated with the example of the Barents Sea shelf ..

1. Unique region

The Arctic is rightly considered to be a unique region of our planet, because, despite its severe climatic conditions and geographical location, this region possesses virtually all kinds of natural resources. The Arctic has large reserves of energy resources, with 90 billion barrels in potential oil reserves, 47.3 trillion cubic meters in potential gas reserves and 44 billion barrels in potential gas condensate [1]. The Arctic marine environment is the habitat of many unique animal species, including particularly rare ones such as the polar bear, the narwhal, the walrus and the beluga. More than 150 species of fish inhabit Arctic and subarctic waters, including cod and flounder, important for the fishing industry [1].

The Arctic harsh environment is particularly vulnerable, which means that the region faces the issues of maximum preservation of the natural environment, development and implementation of a rational "environmentally balanced model of natural resources' sustainable management". The economic interest in using the Arctic resources is steadily increasing. Intensified economic activity in the Arctic upsets the ecological balance. The Arctic ecosystem is quite sensitive to anthropogenic impact and takes very long to recover after "mindless" intervention. In addition, the climatic and hydrological parameters of the Arctic Ocean such as depth, salinity, temperature, velocity and direction of currents, stratification of water and total water balance contribute to a significant dilution of polluted effluents and to heavy deposition of pollutants that remain in marine ecosystems for a long time.

The conditions in which the shelf is developed are difficult, which is why a special focus has to be given to the environment, with each of the development stages having its own specifics. Much effort is currently put into creating a comprehensive environmental protection policy for the region, where not only the countries located along the Arctic coast, but also international organizations and indigenous peoples participate in the decision-making process. Arctic geopolitics is changing [2].

The issues concerning the development of the oil and gas industry and its interaction with the environment are of particular importance, due to significant anthropogenic impact that the industrial development of hydrocarbons on the Arctic shelf has on the environment. This impact is made up of oil spills, greenhouse gas emissions, release of contaminants into the atmosphere, and other factors.

2. The Russian Arctic

Russia has a special geopolitical position in the Arctic region. Environmental safety is crucial for sustainable development of the Russian Arctic. Monitoring of economic activity on the shelf is necessary during active development of hydrocarbon resources of the North and their transportation by sea: this would prevent possible emergencies of anthropogenic nature.

At a major annual press conference held in Moscow on December 14, 2017, the President of the Russian Federation Vladimir Putin discussed the development of the Arctic region. "Industrial development of the Arctic, development of mineral reserves should go together with protection of nature," he stated, answering a journalist's question about the priorities of the state policy in the Arctic.

The current trend in the development of the Russian Arctic is towards intensified economic use of the region's potential. The task is to update the transport and military infrastructure. [3]

The Russian Arctic region is divided into the western and eastern parts along the eastern border of the Vilkitsky Strait (along 105°54′E). The Western Arctic region combines the components of the continent-ocean profile of three oil and gas provinces: the Barents-Kara basin, also called the oil and gas province of the Arctic seas, along the ocean margin, and the Timan-Pechora basin and six Arctic oil and gas regions of the West Siberian oil and gas province (South Kara, Yamal, Gydan, Ust-Yenisei, Nadym-Pur and Pur-Taz) along the continental margin.

This combination of traits is particularly obvious in the basins of the Kara and Barents seas which are some of the most productive shelf seas, with significant deposits of hydrocarbons found on their shelf. At the same time, since the development of energy resources is the key factor governing the social and economic status of the Arctic region, the areas with the largest gas reserves located in oil and gas regions should be taken into account when determining the region's boundaries. In addition, the Barents Sea is known for its high biological productivity, equal to that of the North Sea. Among the thirteen shelf seas, it is these two seas that have both high productivity and multiple resources and can be classified as large marine ecosystems.

3. The environmental conditions in the Barents Sea

Economic efficiency and environmental safety are the main conditions for successful development of oil and gas resources on the Arctic shelf of Russia [4]. The second condition is very important for Arctic seas because the cold arctic climate renders marine ecosystems particularly vulnerable due to ice, slow wave movement, low chemical and biological activity [5]. Biodiversity (of species) is one of the key indicators of the ecosystem sustainability .

Marine biota of the Barents Sea contains a great variety of phytoplankton, zooplankton, benthos, ichthyofauna, avifauna and marine mammals.

To date, unicellular algae of the pelagic zone of the Barents Sea have been rather well-studied, with more than 300 valid species of phytoplanktonic algae found. These algae belong to 8 taxonomic groups: diatoms or Diatomeae (Bacillariophyta), dinoflagellates or Peridineae (Dinophyta), green algae (Chlorophyta), Haptophyta, Prasinophyta, Cryptophyta, golden algae (Chrysophyta) and euglenids (Euglenophyta). The majority of the recorded species are diatoms and dinophytes (about 150 species (50%) and about 120 species (40%), respectively). The share of the remaining species is small and does not exceed 10% [6,7,8]. The abundance of phytoplankton in the Barents Sea ranges from 0.02 thousand cells/l in winter to 2,000 thousand cells/l at the peak of spring bloom. The phytoplankton biomass changes throughout the year: from 0.14-0.4 μg/l in winter and autumn to 3000 μg/l in the spring maximum [7]. The average annual values of primary production vary from 20-25 gC/m²/yr in the northern regions to 200-250 gC/m²/yr in the south-western part of the Barents Sea [5].

While much data has been obtained on the zooplankton of the Barents Sea in general, the zooplankton of some northern regions (especially the northern sector of the sea) remains poorly studied. The list of zooplankton taxa of the Barents Sea [8] includes about 200 species of animals. The zooplankton is divided into taxonomically close groups. A separate large group consists of unicellular zooplankton. Multicellular zooplankton is represented by holoplankton whose entire life cycle is spent in the pelagic zone: Coelenterata, Ctenophora, Crustacea, Gastropoda, Chaetognatha, Rotatoria, Appendicularia and meroplankton (pelagic larvae of benthic animals). The average mass of zooplankton in the upper 100-meter layer of the Barents Sea is more than 200 mg/m³. Areas of increased concentration (more than 500 mg/m³) are found in the waters of the continental shelf. The biomass of zooplankton in the cold Arctic regions does not exceed 100-200 mg/m³. Only a small group in the diverse zooplankton species of the Barents Sea plays a significant role in forming the biomass of the community and, accordingly, takes a prominent place in the diet of fish, mainly crustaceans; these are, mainly, copepods, euphausiids and hyperiids.

About 2300 species of invertebrates, including benthic species, live in the Barents Sea [9]. The fauna of the Arctic seas typically has the following ratio of benthic invertebrate groups: macrobenthos (organisms that are visible to the naked eye) makes up about 60% of the total number of species, meiobenthos (organisms with the size less than 1 mm) makes up 34%, and plankton 6% (planktonic invertebrates are represented by 16 species of radiolarians) [9].

In the entire history of observations of ichthyofauna in the Barents Sea under different climatic conditions, 182 species and subspecies of fish belonging to 59 families, 28 orders and 5 classes were found [5]. However, according to expert estimates, no more than 126 species and subspecies of fish that either constantly inhabit the Barents Sea or spend a part of their life cycle can be encountered in actual surveys [5]. No more than 50 species and subspecies of fish can be simultaneously encountered in the central region of the sea. Other species pass through a certain region in certain seasons, temporarily, during feeding or spawning migrations, or are observed only sporadically. The main part of the ichthyofauna (103 species, or 56.6%) is confined to demersal or benthic habitats. Most species are omnivores (i.e., able to subsist on a variety of foods) but the main trophic group is made up by benthivores (i.e., organisms eating benthos), comprising 52% [5]. Pelagic species include capelin, Atlantic herring, Arctic cod and pollock; benthic species include cod, haddock, rockfish, halibut, catfish and flounder. The sea lamprey (*Petromyzon marinus*) is the only species living in the Barents Sea that is included in the Red Data Book of Russia (2000). Changes in the behavior of individual fish species have been observed, for example, Atlantic cod (*Gadus morhua*) leaves the continental shelf waters and migrates to deeper waters [11]. These changes in migratory behaviors have been linked with environmental conditions due to physical factors or competition for food. According to the data of [12], climate change in the Arctic has a lesser effect on the health and population of commercial fish than the 'fishing pressure'.

The diversity and abundance of the bird species that regularly occur in open areas of the Barents Sea are low and markedly different in different areas of the sea. The species regularly observed in the Barents Sea included 19 species of marine colonial birds, 9 species of Anseriformes and 9 species of waders [10]. The 19 species found directly in open water areas belong to 4 orders: Gagariformes, Anseriformes, Procellariiformes and Charadriiformes. Laridae are the most widespread birds in the Barents Sea (9 species), including 4 species of skuas. However, the avifauna of the open areas of the eastern part of the sea consists of only three species of sea birds: the Arctic fulmar (*Fulmarus glacialis*), the black-legged kittiwake (*Rissa tridactyla*), and the thick-billed murre (*Uria lomvia*).

Both boreal and arctic species can be found in the Barents Sea as it is strongly influenced by the North Atlantic Current in most parts of its waters. The most numerous species of Cetacea are the minke whale, the fin whale, the humpback whale, the killer whale, the Atlantic white-sided dolphin and the white-beaked dolphin. The Arctic fauna consists of pagophilic species inhabiting the cold Arctic waters and the floating ice zone, such as the polar bear, the Atlantic walrus, the bowhead whale, the beluga whale, the narwhal, the harp seal, the ringed seal and the bearded seal.

Species diversity can be affected not only by climate change but also by changes in the environment due to the anthropogenic impact, such as urban or industrial runoff and emission, the products of hydrocarbon exploration, extraction and refinery, heavy metals and other metallurgical waste, pollutants from coastal military bases, waste from vessels operating on nuclear fuel.

The Arctic is attracting growing attention of entrepreneurs in many countries, including Russia; in view of the above, the biodiversity of regional marine ecosystems must be protected as effectively as possible. Ref. [13] proposed to create a network of protected marine areas, selecting them using a methodology based on the MARXAN decision support tool, supplemented by extensive post-analysis. Determining such priority areas for conservation is extremely important.

According to PINRO data (Nikolai M. Knipovich Polar Research Institute of Marine Fisheries and Oceanography) of February 2012, the specific activity of the artificial radionuclide ^{137}Cs was 0.60 Bq/kg dry weight in the bottom sediments of the Barents Sea, and that of ^{90}Sr was 0.85, which corresponded to the background values from the literature data. The artificial background radiation in the coastal waters is generated by objects posing a radiation hazard, concentrated in the Kola and Motovsky Bays, in the cities of Murmansk, Severomorsk, Polyarny, Gadzhiyevo, and in the Sayda, Olenya, Pala, Zapadnaya Litsa, Ura and Ara bays. The water masses of the Barents Sea had a relatively high activity of ^{137}Cs (up to 40-90 Bq/m^3) in the 1980s but by the mid-1990s its content dramatically decreased (to about 15 Bq/m^3). The activity of artificial radionuclides continued to decline in the 2000s. The volumetric activity of ^{137}Cs in the sea water varied from Bq/m^3 in 2005-2008. The activity was distributed relatively evenly in the open sea waters, amounting to 1-2 Bq/m^3 but the activity of ^{137}Cs is higher in the coastal zone of the Kola Peninsula in open bays (Zelenetskaya Guba, Teriberskaya Guba and others) and amounts to 3.0-3.6 Bq/m^3.

According to MMBI (Murmansk Marine Biological Institute), the total content of petroleum hydrocarbons (aliphatic and polyaromatic) in the upper water layer of the Barents Sea has been varying from 0 to 1.13 mg/l for several years. According to PINRO data, in different seasons of 2010, the concentrations of aliphatic hydrocarbons (n-paraffins) varied from 0.84 to 6.6 μg/l in the surface water layer and from 0.56 to 4.3 μg/l in the benthic layer, and did not exceed the threshold concentration of 50 μg/l established for the fishing industry.

Technogenic and oil pollution of the environment is assessed by the content of polycyclic aromatic hydrocarbons (PAHs). Their total concentrations in the waters of the Barents Sea are generally low, and the composition is poor. According to PINRO data, in different seasons of 2010, the total PAH concentration in the water of the Barents Sea varied from 0.96 to 122 ng/l in the surface water layer, and from 0.41 to 144 ng/l in the benthic layer, exceeding the global background level of 20 ng/l. Higher concentrations of PAHs are characteristic for the Atlantic and Barents Sea waters. The total concentrations of carcinogenic compounds (benz[a]anthracene, benzo[b]fluoranthene, benzo[k]fluoranthene, benzo[a]pyrene, indeno[l,2,3-cd)]pyrene and dibenz[a,h]anthracene) varied from 0.03 to 6.3 ng/l in the surface water layer and from 0.03 to 5.7 ng/l in the benthic layer, generally not exceeding 1.5% of the total PAH concentration. Naphthalene, 2-methylnaphthalene, 1-methylnaphthalene, phenanthrene, which are the products of combustion of fossil fuels and organic raw materials, were prevalent in the pelagic layer.

The total concentrations of isomers of hexachlorocyclohexane in the surface water layer of the Barents Sea varied from 0.16 to 5.8 ng/l and from 0.15 to 5.7 ng/l in the benthic layer in 2010. The total concentrations of DDT metabolites varied from 0.16 to 1.8 ng/l in the surface water layer and from 0.12 to 1.4 ng/l in the benthic layer.

The concentration of polychlorinated biphenyls in the open water area of the Barents Sea is very low and is detected by highly sensitive methods of analysis in quantities not exceeding several nanograms per liter of water or in trace amounts. The average concentration of these compounds is 4.29 ng/l according to MMBI data. The total concentrations of detectable organochlorine pesticides and polychlorinated biphenyls in the surface and benthic layers did not exceed the threshold concentration of 10 ng/l at all stations.

The concentrations of detergents (surfactants) were insignificant in the coastal waters as well as in the open areas of the Barents Sea; an increase was occasionally observed in the waters of the Motovsky Bay.

Pollutants in the form of aerosols are released into the Barents Sea from the territory of Northern Europe and the Kola Peninsula; this gets especially acute in winter. Aerosols in the atmosphere of the Barents Sea sub region are considerably saturated with heavy metals (HMs) from regional sources (Ni, Cu, Zn, Cd, Pb, Cr and Hg). The mass of heavy metals released into the Barents Sea with precipitation over the year is by far lower than the mass of the HMs already contained in its waters. Therefore, the release of HMs with precipitation does not lead to noticeable changes in the sea water composition.

According to PINRO data, the concentrations of 11 heavy metals and trace elements (copper, zinc, nickel, chromium, manganese, cobalt, lead, iron, cadmium, arsenic and mercury) examined in different seasons of 2010 in the waters of the Barents Sea tended to increase from west to east and were significantly lower than the threshold concentrations established for the fishing industry. The concentrations of lead varied from 0.10 to 1.4 μg/l in the surface water layer, and from 0.10 to 1.3 μg/l in benthic layer, the concentrations of cadmium varied from 0.01 to 1.3 μg/l in the surface layer, and from 0.01 to 0.90 μg/l in the benthic layer, the concentrations of arsenic varied from 0.10 to 0.80 μg/l in the surface layer, and from 0.10 to 0.80 μg/l in the benthic layer. The concentrations of mercury in the water column varied from analytical zero to the threshold concentration of 0.10 μg/l.

As the pollution of bottom sediments is fairly stable over time in contrast with the pollution of water masses, it follows from publicly available data that elevated levels of bottom sediment pollution are observed mainly in the eastern part of the Barents Sea, as well as in the coastal zone of the Western Murman.

According to MMBI data, the concentration of oil products in the bottom sediments of the open areas of the Barents Sea is very small and varies in the range from trace values to 80 μg/g of dry sediment. The highest concentrations were recorded in the sediments of the Central Groove and the coastal zone. According to PINRO data, in 2010 the content of aliphatic hydrocarbons in the upper 1-3 cm layer of the bottom sediments of the Barents Sea varied from 0.19 to 6.8 μg/g dry weight. Hydrocarbons of both biogenic and petroleum origin were detected in the composition of aliphatic compounds of most samples, as indicated by the dynamics of the pristane/phytane ratio > 1 or ≤ 1. Russian Federation has not adopted any standards for the n-paraffin content in bottom sediments. The total content of n-paraffins in the bottom sediments of the examined areas of the Barents Sea was significantly lower than the technogenic background level typical for the upper layer of the bottom sediments of the Western Arctic Shelf, amounting to 340 μg/g dry weight.

The publicly available data indicate a relatively low content of polycyclic aromatic hydrocarbons in the bottom sediments of the Barents Sea. Russian Federation has not adopted any standards for PAH content in sea bottom sediments. According to the classification of pollution levels of sea bottom sediments adopted by the Norwegian Pollution Control Authority (SFT), the content of PAH and benz[a]pyrene in the bottom sediments did not exceed the background levels < 300 and < 10 ng/g dry weight, respectively, in the examined areas of the Barents Sea.

Residual amounts of α, β, γ-isomers of hexachlorocyclohexane (HCH) and DDT metabolites were the prevalent organochlorine pesticides in the bottom sediments of the Barents Sea examined in 2010. The total content of DDT in bottom sediments varied from 0.37 to 7.3 ng/g dry weight at all stations and exceeded the technogenic background level of 0.5 ng/g dry weight, which corresponded to the “moderately polluted” and “markedly contaminated” categories.

According to the PINRO studies, in 2010 the PCB content in the bottom sediments of the Barents Sea varied from 0.67 to 2.1 ng/g dry weight and did not exceed the technogenic background level of 5 ng/g dry weight.

According to the MMBI data, the level of PCB accumulation in the bottom sediments of the Barents Sea can also be categorized as very low. The concentration of these compounds in 35% of the tested precipitation were below the detection threshold. The concentration of PCBs in the sediments of open sea areas averaged 0.3 ng/g dry weight. A relatively PCB high content (1.0-1.5 ng/g dry weight) was detected in the sediments of the Central Groove (Polar Front region) as well as in the coastal zone.

Most of the heavy metals and trace elements, except for iron and manganese, do not accumulate in the sediments in substantial concentrations. According to MMBI data, the spatial distribution of copper, cadmium, nickel, chromium, iron, manganese, zinc and arsenic has a general trend towards the concentrations increasing from coastal zones to the deep water regions of the Central Bank. Russian Federation has not adopted any standards for the content of heavy metals in sea bottom sediments. According to the pollution criteriain sea sediments adopted in Norway for arsenic content, all the bottom sediments examined can be classified as “moderately polluted”. The cadmium content in bottom sediments corresponded to the “moderately polluted” and “markedly contaminated” categories at most stations. The content of the remaining metals corresponded to the natural background levels.

The results presented confirm that the present pollution level of the marine environment of the Barents Sea is relatively low, with the exception of the narrow coastal zone of the Kola Peninsula.

4. Environmental risks in developing energy resources

The studies based on modern methods of environmental monitoring have revealed no significant changes in the current environmental status of the Arctic region. However, the existing methods of medium and long-term assessment of marine biota habitats are rather imperfect. It is known that Russian rivers annually carry up to 500,000 tons of oil products (including those formed in accidents during oil extraction and transportation) into the Arctic Ocean. This means that the negative environmental impact of developing energy resources in the Arctic region is likely to result in an irreversible environmental crisis.

It should be understood that intensified search and extraction of hydrocarbons in the Arctic will lead to increased pressure on the Arctic ecosystems. With no effective mechanisms of environmental control, this could exacerbate the environmental problems even more, especially on the continental shelf of the Barents, Pechora and Kara Seas.

Environmental hazards in the extraction of energy resources in the Arctic include drilling wells, which can cause pollutants’ release into the atmosphere and into the marine environment, volatile emissions of liquid hydrocarbons and hydrocarbon gases from the wells during drilling, emissions of greenhouse gases that can lead to climate change, transportation of hydrocarbons by tankers and pipelines, which carries a risk of spills during loading /unloading and emergencies.

Even insignificant spills of extracted energy resources lead to irreparable environmental damage, especially on the shelf that is covered by thick ice for most of the year.

5. Acknowledgment

The research was carried out in accordance with the basic part of the state assignment to the higher educational institutions of the Ministry of Education and Science of Russia in the part of the initiative research projects on the topic of R & D "The rational organization of economic development and the maritime transportation of energy resources in the Russian Arctic", No.13.12713.2017 / 8.9.

References

[1] Miloslavskiy V 2017 *Young scientist* B 18 157
[2] Knecht S and Keil K 2013 *Polar Journal* B **3**(1) 178 DOI: 10.1080/2154896X.2013.783276

[3] Frolov I E 2015 *Studies on Russian Economic Development* B **26(6)** 561 DOI: 10.1134/S1075700715060040
[4] Amiragian A 2016 Oil and gas in the Russian Arctic *FEC of Russia* 9 34
[5] Matishov G 2011 *Comprehensive studies of large marine ecosystems* p 516
[6] Kuznecov L and Shoshina E 2003 Phytocenoses of the Barents Sea (physiological and structural characteristics) p 308
[7] Makarevich P and Drugkova E E 2010 *Seasonal cyclic processes in coastal planktonic algalocenoses of the northern seas (*Federal State Budgetary Institution of Science Southern Scientific Center of the Russian Academy of Sciences) p 338
[8] Matishov G et al 2000 Biological Atlas of the Arctic Seas 2000: Plankton of the Barents and Kara Sea *National Oceanographic Data Center NOAA Silver Spring MD USA* p 356
[9] List of species of free-living invertebrates of Eurasian arctic seas and adjacent deep waters: Exploration of the Fauna of the Seas 2001 B **51(59)** p 133.
[10] Khfsnov Yu and Nikolaeva N 1996 *Ecosystem of pelagic seas of the Western Arctic* 101
[11] Ingvaldsen R B et al 2017 *Polar Biology* B **40** (10) 2105
[12] Durant J M and Hjermann D O 2017 *Marine ecology progress series* B **577** 177
[13] Solovyev B et al 2016 Hawaii World Conservation Congress Honolulu, HI published in AQUATIC CONSERVATION-MARINE AND FRESHWATER ECOSYSTEMS B **27** (1) 30

IOP Conf. Series: Earth and Environmental Science **180** (2018) 012008 doi:10.1088/1755-1315/180/1/012008

The role of the Northern Sea Route in Russian LNG Projects Development

E G Katysheva

St. Petersburg Mining University, St. Petersburg, Russia

helene_la_belle@mail.ru

Abstract. Currently, there has been a significant increase in the volume of freight traffic on the Northern Sea Route (NSR). The following advantages of the Northern Sea Route can be mentioned: no piracy in the region, less time taken for cargo transport, ship bunkering with natural gas from onshore or offshore fields. The paper indicates that the requirements for the bunker fuel quality and the amount of pollutant emissions into the atmosphere during shipping in the Arctic zone are very stringent. This work states that one of the most important ways to follow the emission limits is changing the fuel of the ice-rated vessels for liquefied natural gas (LNG). It is noted that Russia has a number of opportunities to apply LNG both to supply the Arctic Region with water and surface transport and to supply the onshore settlements and enterprises. It is noted that the use of gas as fuel will decrease air pollutant emissions. Furthermore, shift to LNG as bunker fuel will resolve a major environmental problem of accidental spillage of marine fuel and crude oil. The article states that LNG is a reliable and competitive fuel for shipping in the Arctic Region. It is indicated that the growth of the Russian domestic LNG market in the form of industrial consumers and ships bunkering can be an incentive for the intensive domestic LNG-projects development.

1. Introduction

The Northern Sea Route (NSR) is the shortest sea way between Europe and Asia. It goes through the seas of the Arctic Ocean (the Kara Sea, the Laptev Sea, the East Siberian Sea, and the Chukchi Sea) for around 2,500 nautical miles. On an average, passing along the Northern Sea Route will take 10.6 of a day [7]. Due to the development of new industrial projects in the Arctic Region, a significant increase in traffic volume along the Northern Sea Route can be currently observed (table 1).

To protect the environment against polluting emissions from shipping, now, requirements to the quality of marine fuel are being updated and tightened, new environmental regulations worked out [1]. The strictest regulations are applied to some air pollutant Emission Control Areas (ECA),. In Europe, ECA include Baltic Sea and Northern Sea areas, in the Northern America – Pacific and Atlantic coast areas.

Tight restrictions applied to air pollutant emissions (i.e. soot, nitrogen oxides and sulfur oxides) in use of marine fuel in ECA zones adopted by the International Maritime Organization can be implemented by taking the following measures:

- use of low-sulfur oil fuel of higher quality;
- installation of treatment facilities;
- liquefied natural gas preference (LNG).

Table 1. Traffic Volume Forecast along the Northern Sea Route until 2030, mln tons

Cargo	Fact	Forecasted		
	2016	2020	2025	2030
Hydrocarbons, total	8.7	30.7	50.8	69.4
including:				
oil and gas condensate	8.7	19.7	31.7	36.7
LNG	0.0	11.0	19.1	32.7
Ore and concentrated ore	0.7	1.0	1.0	1.0
Carbon	1.0	8.0	20.0	30.0
Northern Supply	1.0	1.5	2.0	2.0
Transit	0.0	1.0	1.2	2.0
TOTAL	11.4	42.2	75.0	104.4

Source: [7].

Possessing considerable resources of natural gas in the Arctic Region and investing heavily in LNG production in the region, Russia has all the conditions to use this fuel both for supply of water and surface transport in the Arctic Region, and for supply of onshore settlements and enterprises.

2. Research questions

According to the data from the Russian Registry of Shipping, today, in the ports of the Arctic Region of Russia, 439 ships of all rates are recorded [8]. Of these, 74% use diesel fuel, 1% are running on oil residue, and 25% - on motor and diesel oil. This ratio will make shifting to low-sulfur fuels quite painless, without significant change in the fuel system of a ship. Furthermore, age class composition of the fleet, and the fuels used can be deemed conducive to the shifting to LNG of some vessels. At the present time, it seems technologically viable to shift all kinds of vessels operating in the Arctic Region of Russia. In addition, there are projects for use of LNG within the process of large-scale fleet upgrade and ship building for new industrial projects in the Arctic Region.

The forecasted increase in cargo carriage in the Arctic Region is primarily determined by exploration of hydrocarbon resources in the Timan-Pechora province, Yamal, and Taimyr. Carriage of equipment for field development in the Arctic Region and the necessity to remove the found raw material will require the establishment of a large-scale and complex infrastructure for delivery and storage of oil fuels. Alternatively, the Arctic LNG projects can be integrated in transportation and energy generation patterns of already existing ones [5]. It should be emphasized that implementation of the projects for the construction of industrial and extracting facilities in the Arctic region will inevitably result in significant increase in cargo carriage volume and aggravating of negative impact on the environment by the ships delivering materials and equipment and removing finished products.

The role of the Arctic Region as a transportation corridor is currently becoming more prominent due to reduced ice cover in some Arctic basins. Therefore, the Arctic Region of Russia can be considered an important transport route bridging Europe and Asia. The main advantages of the Northern Sea Route include the absence of piracy in the region and cutting time for carriage. For example, cargo carriage from Norway to Japan may take up to 21 days less than through the Suez Canal [9].

Moreover, the Northern Sea Route provides ships with a possibility to bunker with natural gas from both onshore and offshore fields. This will also promote competitive advantage of the Northern Sea Route [6].

In the Arctic Ocean area, 3 major projects on LNG production are currently being implemented. Yamal LNG, Pechora LNG, and Arctic LNG-2. It is viable to arrange bunker centers in the vicinity of LNG production plants. Also, storage and bunkering facilities in the ports of Murmansk, Archangelsk, Dikson, Tiksi, and Anadyr will be required.

Directions and prospects for LNG application as bunker fuel depend on the transportation geography. In the Arctic Region of Russia, ships are designed for shuttle runs along permanent routes, while their high ice rate enables them to run during the entire term of operation.

If compared with oil fuels, LNG has the superior calorific value and the best physical characteristic (Table 1). However, a fuel system of greater capacity is required onboard because of the lower density of LNG.

Table 2. Physical properties of different fuels

Physical properties	Gasoil	Residual oil	LNG
Density, kg/m^3	860 – 900	920 – 1010	422
Net calorific value, MJ/kg	41,4 – 43,3	40,5 – 41,4	50,02
Boiling point, ^{0}C	180 – 360	x	- 162
Self-ignition point, ^{0}C	250 – 300	x	595
Lower explosion level, vol.%	0,5	x	4,4
Upper explosion level, vol.%	7,5	x	15 –17

Source: [7], data from companies

However, it should be noted that the existing marine fuel system capacity for oil fuels is sufficient to make shuttle runs on LNG in the Arctic Region even in ice-bound conditions. Availability of LNG for carriage gurantees ship bunkering in the Yamal semi-peninsular region, which does not require increased capacity of a fuel system, nor it leads to cargo capacity loss. In the absence of ice, LNG stored in oil fuel tanks can procure passage along the route Dudinka – Murmansk – Dudinka.

Therefore, LNG is a reliable and competitive fuel for shipping in the Arctic Region. With increase of shipping and routes length, efficiency of LNG application also significantly increases. The payback period for LNG is from 2.5 to 5.5 years [2].

Using gas as fuel favors reduction of air emissions. The most notable reduction is observed with SOx, i.e. particulates. NOx emissions will decrease by 80% [1]. In addition, greenhouse gases emissions are reduced. Environmental safety of LNG usage is primarily determined by modes and kinds of ship engines. Engines operating on gas as well as two-fuel modified engines allow for methane emissions. Making improvements in gas engines will eliminate this drawback.

Shifting to LNG as bunker fuel will assist in resolution of the acutest environmental problem of accidental spillage of marine fuel and crude oil. Oil spillage has negative impact on all food chain elements in the Arctic Region. Some representatives of the Arctic fauna are particularly sensitive to oil spillage, since pollution with oil and oil products destroys fur and feather heat insulation properties. Currently, there is no reliable technique for elimination of accidents of this kind under the ice-bound conditions. The problem is very acute for polar basins where accident management activities are complicated not only by thick ice coverage but also by polar night, low temperature, strong winds, and lack of infrastructure facilities.

Experts estimate that the expenses for elimination of effects of residual oil spillage of 100 tons in the Arctic region will bring to naught all price advantages from use of this cheap yet environmentally hazardous fuel by all ships of the global Arctic fleet for three years [10]. In this respect, in 2011, resolution adopted by the International Maritime Organization took effect. It prohibits use and transportation of crude oil with the density exceeding 900 kg/m^3, heavy oil products, tars, and pitch in the Antarctic waters. Similar restrictions seem desirable for the Arctic Region as well.

Obviously, for some segments of shipping, the transition to LNG will be in demand. Given the political will and positioning of the Northern Sea Route as the world's first sea transport corridor for LNG, the opportunities and scope of LNG using will be maximally realized. At the same time, Russian gas resources will reach a new promising and rapidly developing segment in the world - bunker fuel.

In fact, absolute compliance with existing and planned environmental requirements ensures the attractiveness of using LNG in the long run. The high economic and environmental efficiency of LNG using can ensure the dominance of LNG as bunker fuel in the Arctic zone.

If the existing fleet does not require significant investments in its infrastructure and there is a possibility of supplying low-sulfur fuels in the required volumes, with the expected increase in cargo turnover and, accordingly, fuel requirements of up to 5-6 million tons, there are justifiable doubts about the ability to provide ships with fuel in the required volume and at competitive prices.

The qualitative assessment of LNG used on new vessels shows increased capital costs, certain capacity limitations due to the need to install volumetric fuel tanks and lower operating costs. Individual estimates show that higher capital investments in LNG vessels result in a slight (1-2%) increase in the cost of transportation [4].

The main advantages of LNG in the Arctic zone will be its availability in the long run and compliance with current and future environmental requirements.

The creation of LNG infrastructure for bunkering ships will create conditions for the social development of the Arctic ports areas.

Data on the cost of ships on LNG and the transfer of vessels to LNG use are very limited due to the limited number of vessels on LNG. The available estimates provide the following information on additional capital investments in ships for the LNG using:

- for the dry cargo ship of the ice class with a deadweight of 70 thousand tons and a propulsion system of 40 MW, the additional costs amount to $ 26.7 million;
- for a tanker with a propulsion system of 12 MW and a deadweight of 75 thousand tons, additional investments of $ 9.5 million are required [7].

The Russian Arctic zone requires the use of powerful ships of high ice class. However, the cost difference is solely related to the fuel system and engine of the ship, therefore for economic calculations it is possible to use a range of additional capital investments from $ 9.5 to $ 26.7 million for vessels with propulsion systems of 12 and 40 MW or $ 790-670 per kW, respectively [5].

With the actual absence of modern infrastructure for storing and distributing oil fuels in the Arctic, for the purposes of this analysis one can accept the thesis of equal capital costs for creating an infrastructure for LNG and for oil fuels, with the expected significant increase in cargo turnover. Thus, the difference in capital costs will be concentrated only in the difference in the acquisition of ships working on LNG and oil fuels.

The main attraction of LNG usage, in addition to the fact this type of fuel has to meet modern and future environmental requirements, is its relatively low price. Over the past 6 years LNG has had significantly lower energy unit prices compared to petroleum fuels, and this creates additional positive effects from the use of LNG in addition to environmental benefits.

As a rule, the main LNG consumption centers are located at a considerable distance from the production centers, so the transportation of LNG to the consumer adds to its cost.

Due to the use of new world-class deposits and environmental cold Russian projects in the Arctic make it possible to obtain LNG at low prices.

3. Conclusion

Liquefied natural gas is a new fuel to be used for marine ship bunkering. In the view of tough requirements to the pollutant emissions from ship engines running on oil fuel, and price advantage of

LNG, liquefied natural gas can be regarded as a fuel that can come to dominate the marketplace of marine fuel.

The problem of extending usage of LNG as marine fuel is particularly acute for Russia possessing considerable gas reserves in the Arctic Region, and liquefying production facilities. Shifting to LNG will require essential upgrade of the vessels operating in the region. However, the problem must be solved anyway, because of the high degree of the Arctic fleet wear in Russia. Construction of new ships designed for LNG seems to be the first choice. Alternatively, re-equipment of the existing ships can be considered.

LNG will reduce the risk of spillage of oil and marine fuel, and diminish air pollutant emissions. Extensive activities in the area will strengthen positions of the Russian manufacturers on LNG-based fuel market promoting implementation of advanced construction technologies for ice-rated ships.

Use of LNG as bunker fuel will give rise to an opportunity to supply industrial facilities and settlements of the Arctic Region with LNG-based energy. Combined use of LNG infrastructure for bunkering to supply electric power plants and heat generating facilities with gas, will reduce operating expenses and set favorable prices for all classes of customers.

Implementation of major LNG production projects in the development of transit carriage along the Northern Sea Route will increase the LNG consumption rate significantly. Growing competition on LNG marketplace necessitates the search for new segments of this market. The LNG domestic market growth represented by industrial consumers and ship bunkering may initiate intensive development of the domestic LNG projects.

It is evident that the use of LNG as fuel will reduce the environmental impact, including, but not limited to COx, NOx, SOx emissions.

Implementation of LNG production projects in the Arctic Region such as Yamal LNG, Pechora LNG, Arctic LNG will strengthen Russian position in the global LNG market, and form a transportation corridor between Europe and Asia for exclusive LNG bunkering.

Availability of LNG and reliability of supplies from different production facilities may ensure essential price advantage of LNG as compared with other routes. Thus, LNG cargo carriage may significantly increase.

Intensive exploration of the Arctic Region is associated with growing supplies of marine fuel, as well as with the establishment of infrastructure for transshipment, storage and bunkering for the purpose of transit and removal of oil, carbon, and gas, from scratch. Under these conditions, extensive substitution of oil fuels with LNG is of immediate interest for Russia. It will reduce costs for carriage and transit, and protect the delicate Arctic ecosystem from pollutant emissions and effects of spillage of oil and oil products.

References

[1] Ametistova L and Knizhnikov A 2017 *Environmental aspects of LNG-projects in arctic conditions* (Moscow) (in Russian)

[2] Ampilov Y 2018 A new look at Arctic prospecting and exploration *Oil and Gas Russia* **5** (115) 72-78 (in Russian)

[3] Balashova E S and Gromova E A 2017 Arctic shelf development as a driver of the progress of the Russian energy system *MATEC Web of Conferences 23 May 2017* **106** 06008 (in Russian)

[4] Borisov V N and Pochukaeva O V 2017 Investment and innovative technological efficiency *Case study of the Arctic project Studies on Russian Economic Development* **28** (2) 169-179 (in Russian)

[5] Katysheva E and Tsvetkova A 2017 The future of oil and gas fields development on the arctic shelf of Russia *Conf. Proc. of International Multidisciplinary Scientific GeoConf. Surveying Geology and Mining Ecology Management, SGEM 17* **53** 935-940

[6] Katysheva E G 2017 Localization of Equipment Production for the Oil Industry in the Russian

Arctic *RRI-2016 European Proc. of Behavioural Sciences* **26** 387-392
[7] Klementiev A, Knizhnikov A and Grigoriev A 2017 *Prospects and possibilities of LNG using for bunkering in the Arctic regions of Russia* (Moscow)
[8] Knizhnikov A, Klementiev A and Grigoriev A 2017 Insights in the Arctic Region *Oil and Gas Russia* **5** (115) 72-78
[9] Komkov N I, Selin V S, Tsukerman V A and Goryachevskaya E S 2016 Scenario forecast of the development of the Northern Sea Route *Studies on Russian Economic Development* **27** (2) 180-188
[10] Komkov N I, Selin V S, Tsukerman V A and Goryachevskaya E S 2017 Problems and perspectives of innovative development of the industrial system in Russian Arctic regions *Studies on Russian Economic Development* **28** (1) 31-38

Maritime economics of the Arctic: legal regulation of environmental monitoring

S Kozmenko [1], A Teslya[2] and S Fedoseev[1]

[1] FIC "Kola Science Center RAS", Apatity, Russia
[2] Peter the Great St. Petersburg Polytechnic University, St. Petersburg, Russia

anntes@list.ru

Abstract. The current favorable natural conditions (melting of the Arctic ice) and the advance of innovative technologies used for extraction and transportation of unique natural resources are reasons for the growing economic interest of the international community in the Arctic region. Intensified economic activity leads to increased anthropogenic pressure on the highly vulnerable Arctic environment. Because of this, a study of conservation and restoration of the Arctic ecosystem is of utmost importance. Increased extraction of resources in the Arctic involves developing the transport system, including sea freight. We have found out that increased interest of the global community in the development of the Arctic led to an increase in the number of countries willing to influence the economic activity in the Arctic, including organizing sea freight. The growing number of participants with their expanding range of competing interests hinders the search for coordinated solutions. So far, the unique legal status of the Arctic has not been established at the international level. The paper confirms that the national legislation of Russia does not offer the full opportunities for forming a system of environmental monitoring either, and, therefore, it does not allow to carry out a full economic assessment of the costs of restoring the Arctic ecosystem. We have proposed the key directions for reorganizing the environmental monitoring system, based on recognition of the unique status of the Arctic; these directions can form the future basis for the assessment procedure concerning marginal social costs and benefits of developing the Arctic spaces and making decisions on the scale and trends of its development.

1. The Artic region

There is currently no unified international legal act establishing the definition of the Arctic region, which engenders certain problems in determining the Arctic boundaries, as well as in creating strategies for the development of the Arctic and environmental protection of its unique land and sea areas. Geographically, the Arctic includes the northern parts of Eurasia and America continents, almost the entire Arctic Ocean with islands, and adjacent parts of the Atlantic and Pacific Oceans. The southern boundary of the Russian Arctic zone practically coincides with the southern boundary of the tundra. The continental part of the Arctic zone of the Russian Federation makes up about 35%, with 2% occupied by islands and archipelagos. Accordingly, the water areas of the Arctic Ocean and inland seas occupy 63% of the territory of the Russian Arctic (65% are taken up by the seas and 35% by the Arctic Ocean). The marginal seas (the Barents, the Kara, the Laptev, the East Siberian, the Chukchi, Beaufort and the Greenland seas, the Baffin Bay, etc.) constitute the shelf region. The United Nations Convention on the

Law of the Sea (UNCLOS) that the Russian Federation signed and ratified in 1997 establishes the total area of the Russian Arctic waters (a 200-nautical-mile exclusive economic zone) to be 4.1 million square kilometers. The Russian Arctic contains all categories of marine areas defined within the UNCLOS: internal waters, territorial sea adjoining the state borders of coastal countries, the contiguous zone, exclusive economic and fishing zones, the continental shelf, the high seas and portions of the seabed areas located outside the continental shelf.

The area of the Arctic region and the size of its population are the subject of substantial controversies. In general (see Table 1), more than 50% of the Arctic territories belong to Russia, with more than 50% of the Arctic population living on these territories. The density of population in the Arctic varies greatly depending on the country, but on average it is low. The density of population on the territory of the Russian Federation also varies considerably. However, in general, according to [1], more than 7% of the world's population live in the Arctic.

Table 1. Territories and populations of the Arctic, 2015

Country, Arctic region	Area, thousand km^2	Percentage	Population, thousands of people	Percentage	Population density, people per km^2
Total	15,189	100%	4,611	100%	0.30
Arctic Zone of the Russian Federation	8,409.3	55%	2,502	54%	0.30
Total in the Arctic territories of the countries bordering the AZ	6,780	45%	2,109	46%	0.31
Canada	2,446.9	16.1%	120	2.6%	0.05
Denmark (Greenland)	2,131.8	14.0%	58	1.3%	0.03
USA	1,718	11.3%	710	15.4%	0.41
Norway	188.6	1.2%	466	10.1%	2.47
Iceland	103.3	0.7%	311	6.7%	3.01
Sweden	98.2	0.6%	260	5.6%	2.65
Finland	92.7	0.6%	184	4.0%	1.98

The climate of the Arctic largely depends on the dynamics of heat transport from the southern latitudes. According to [2], the average temperatures of the Arctic have almost doubled over the past 100 years. The air temperatures are expected to increase at a faster rate up to 2020. The ongoing decline of the multiyear ice in the East Siberian and Chukchi Seas, the reduction in its thickness and quantity contribute to improved navigation along the Northern Sea Route bypassing coastal routes. These climate changes in the Arctic, along with the advances in new technologies, create favorable trends for socio-economic development of the territories. Over time, these trends will have a serious negative impact not only on the Arctic environment, but also on all aspects of life and evolution of the global community [3]. On the other hand, global warming and the rising sea level impose serious threats, both on island countries and on those with long coastlines. Therefore, the strategies of socio-economic development for these territories and the programs of development of sea areas, adopted by the states bordering the Arctic, will have economic, social and environmental consequences for the entire global community [4].

New industrial facilities emerge due to intensified development of natural resources in the Arctic. Due to complex interaction between industrial marine-based systems and natural marine ecosystems, large sectoral and intersectoral maritime clusters are formed in the zone of contact of marine and coastal areas, new types of activity evolve, along with the social and professional composition of the labor force of oceanic industry branches, while the population in the coastal zones grows [5]. At the same time, intensified economic activity has a strongly negative impact on ocean ecosystems and coastal zones in the Arctic [6]. Therefore, it is particularly important to work out the principles and methods for managing the exploitation of natural resources and to generate scientific foundations and practical recommendations for environmental monitoring. Almost all scientific studies related to the economy of the Arctic are aimed at resolving these issues.

The economic interest towards developing the marine space, including the Arctic shelf [7], is largely determined by the relations of the Arctic states with the countries remote from the Arctic [8]. The clash of economic interests in addressing the issues of Arctic development, in particular, using the Northern Sea Route, often arise between strategic partners [9, 10]

While the Arctic countries intend to retain national jurisdiction over sea routes [11], some European Union countries call for considering the Arctic Ocean to be part of the "common heritage of mankind". This is to say that the countries that are geographically remote from the Arctic state seek to use its economic potential and participate in choosing the strategies for development and exploitation of land and sea areas.

Growing interest in the problems of the Arctic means that an increasing number of countries are involved in solving environmental issues of this region, which allows for a more comprehensive analysis of environmental safety, since the Arctic states are not the only source of environmental issues. For example, studies revealed that the most toxic substances are released into the Arctic from sources other than the industrial enterprises located in the region. Ref. [12] identified the following main sources of pollution:

– radioactive materials from nuclear waste reprocessing facilities in the UK and France carried from the North Sea with the waters of the Gulf Stream;
– pesticides carried into the Aral Sea with the effluents from Amu Darya and Syr Darya and with dust storms through the upper atmosphere and the zone of high-altitude jet streams;
– mercury, lead and cadmium transported by high-altitude air flows from Europe, North America, China, South Korea and Japan.

Thus, not only the interaction between the countries bordering the Arctic but also the interaction with the countries geographically remote from the region is needed to ensure environmental safety in the Arctic. Several international organizations were created with the aim of coordinating the development of Arctic land and sea areas. One of these organizations is the Arctic Council, established in 1996 by a joint declaration of Denmark, the USA, Russia, Canada, Norway, Iceland, Sweden and Finland to promote cooperation in the sphere of environmental protection and to achieve sustainable development of the Arctic territories. As the non-Arctic countries' interest in the unique resources of the region grew, such countries as the United Kingdom, France, Germany, the Netherlands, Poland, Spain and China joined the Council as permanent observing states. Thus, the environmental problems of the Arctic have become the problems of the entire global community. On the one hand, this allows finding comprehensive solutions for these problems. On the other hand, however, the expanding range of competing interests in the Arctic, causd by a growing number of participants, significantly increases the economic and transaction costs of finding a possible solution.

2. Legal regulation of environmental safety in the Arctic

The unique international status of the Arctic is not currently defined in any law; this is one of the factors hindering the implementation of a unified environmental monitoring system. International legal regulation of environmental safety in the Arctic is enshrined in the following documents: the UN Convention on the Law of the Sea of 1982, the International Convention for the Prevention of Pollution from Ships of 1973, and the International Convention on Oil Pollution Preparedness, Response and Cooperation of 1990.

All types of marine activity in the Arctic (such as ensuring passage through territorial seas, navigation on the high seas, including the existing 200-nautical-mile exclusive economic zone, exploration and extraction of minerals, fishing, environmental protection, scientific research) are carried out in accordance with the existing conventions. We should stress that the international legal regime of the Arctic region is based on the principles and norms common for the whole World Ocean, and the exclusive status of the Arctic is not established by any international documents. The exclusive status of the Arctic territories is partially regulated by national legislative acts of individual states, or by international agreements on specific activities.

There is also a number of agreements that are not directly related to the Arctic, but are important for environmental control, for example, the 1992 Convention on the Protection of the Marine Environment of the Baltic Sea Area (the Helsinki Convention). As part of intergovernmental cooperation, Russia, Germany, Denmark, Latvia, Finland and other countries of the Baltic Sea participate in the work of Helsinki Commission (HELCOM) coordinating the protection of the marine environment of the Baltic Sea from all sources of pollution.

Important Russian legislation on the subject includes the Federal Law on Environmental Protection of December 26, 2001, the Order on Environmental Impact Assessment of Planned Economic or Other Activities in the Russian Federation (Order No. 372 of the State Committee of the Russian Federation on Environment Protection of May 5, 2000), the Federal Law on the Continental Shelf of the Russian Federation No. 187-FZ of November 30, 1995. These laws establish mandatory environmental impact assessments in order to prevent or mitigate negative impacts and the associated social and economic consequences. State Environmental Expertise aimed at protecting the marine environment and preserving the natural resources has become compulsory for conducting any activity on the continental shelf. State monitoring of the continental shelf is part of the unified system of state environmental monitoring, making regular observations on the marine environment and bottom sediments obligatory. The requirements for the protection of the marine environment in the exploration and extraction of oil and gas of the continental shelf, territorial seas and the coastal zones are established by GOST R 53241-2008 on Marine Geological Exploration.

However, while the legislation is rather extensive, there are no uniform rules regulating the scope, types and methods of environmental monitoring; because of this, the companies extracting the resources tend to use their own assessment methods, developing individual programs for damage assessment and compensation. At present, Russia does not have a single document regulating the entire range of issues related to the development of the Arctic zone and establishing its unique status.

The draft of the Federal Law on the Arctic Zone of the Russian Federation is currently pending approval. Meanwhile, a special approach has to be taken to using and protecting the unique natural resources of the Arctic. So far, the legal regulation of all aspects of the development of the Arctic zone has been based primarily on constitutional, administrative, civil, environmental, land and various other laws. The existing legal norms are incomplete and inconsistent, which hinders the construction of an effective environmental monitoring system and does not allow to fully coordinate the interests of all stakeholders (industrial groups, the state and the population), even within one country.

3. Economic goals and organization of environmental monitoring

Development of the Arctic opens up additional opportunities for intensified extraction of natural resources, which leads to an increase in the volume of traffic (mainly marine), and, consequently, to heavier pressure on the marine ecosystem. The Arctic has vast sea areas, which means that each port involved in local or international trade should minimize the negative impacts on the environment. In this regard, interesting points were raised in [13], proposing the criteria for choosing seaports, which allowed to minimize their number and, accordingly, to relieve the pressure on the marine environment.

An attempt to assess the expediency of developing the sea transport was made in [14], with the conclusion that, due to complex environmental and economic dynamics of the Arctic, close interaction between researchers from various fields of science is needed to carry out comprehensive assessment of all the consequences of developing the territories and to obtain reliable predictions for the future.

The shipping, fishing and travel industries are currently flourishing in a number of Arctic countries. Further development of these sectors should contribute to sustainable socio-economic development of the Arctic [15]. Nevertheless, researchers propose to consider any issues of economic activity in the Arctic not only from the standpoint of economic costs and benefits, but also taking into account the potential impact of the changes on the social sphere, as well as taking into account the pressure on the environment [16].

Some of the most substantial sources of environmental damage to the Arctic territories in Russia include:

- destruction of the active layer in the tundra during construction of roads and pipelines;
- industrial waste disposal and recycling;
- emissions from mining, oil and gas extraction plants into the atmosphere;
- substantial damages due to accidents accompanied by oil spills as a result of depreciation of the main funds, where the consequences cannot be mitigated promptly.

We should note that according to [12], specific emissions from industrial facilities in the Arctic significantly exceed the average emission rates across Russia (Table 2), which means that special standards have to be developed for organizing industrial production.

Table 2. Average specific emissions, % to the average emission rate in Russia

	Russian Arctic	Arctic seas	Zone of atmospheric impact on the Arctic
Number of industrial facilities	69	577	927
Emissions into the atmosphere	1372%	941%	361%
Polluted effluents	208%	176%	-
Land transformation	287%	-	-
Waste production	501%	-	-
Motor vehicle emissions	95%	-	-

Meanwhile, developing the transport infrastructure, organizing the exploration, extraction and transportation of industrial resources are the crucial goals of the Russian Federation regarding socio-economic development of the Arctic region. The priority tasks are developing industrial resource extraction and the transport infrastructure, as well as using the Arctic marine environment while preserving its unique ecosystem or at least minimizing the damage from the development of the region. The volume of cargo traffic is considerably high and remains so throughout the entire year (Table 3).

Table 3. Cargo turnover in the ports of the Arctic Basin (cargo shipped to ports), thousand tons (compiled based on the data available at https://cargo-report.info/stat/ports-arctic/)

Period	Apr 2018	Mar 2018	Feb 2018	Jan 2018	Dec 2017	Nov. 2017	Oct 2017	Sep 2017	Aug 2017	Jul 2017	Jun 2017	May 2017
Murmansk	1747	1631	1582	1705	1968	1746	1998	1725	1670	1558	1735	1800
Kandalaksha	129	194	147	141	158	164	153	181	165	204	126	87
Arkhangelsk	19	79	77	85	88	52	61	52	68	41	63	89
Total	1895	1904	1805	1931	2214	1962	2211	1958	1903	1803	1924	1976

On average, 89% of the total volume of cargo is transported through the port of Murmansk. Notably, the volume of the cargo shipped to the ports considerably (by 50 times on average) exceeds the volume of the cargo shipped from the ports (Table 4). Cargo is shipped mainly from the port of Murmansk, with insignificant amount of cargo shipped from Kandalaksha and Arkhangelsk.

Table 4. Cargo turnover of the ports of the Arctic Basin (cargo shipped from ports), thousand tons (compiled based on the data available at https://cargo-report.info/stat/ports-arctic/)

Period	Apr 2018	Mar 2018	Feb 2018	Jan 2018	Dec 2017	Nov. 2017	Oct 2017	Sep 2017	Aug 2017	Jul 2017	Jun 2017	May 2017
Murmansk	37	32.8	17	35	33	45	32	38	29	33	31	29

It can be seen from assessing the structure of the cargo traffic (Fig. 1) that the majority (more than 70%) of the cargo shipped to the ports is coal; fertilizers, ores and oil cargo account for about 14%, 10%

and 4%, respectively. The rest of the cargo makes up less than 0.3% of the total volume, with 12.04 thousand tons taken up by construction minerals, 0.04 thousand tons — with timber, 0.03 thousand tons with grain, and 0.02 thousand tons with ferrous metals. The cargo shipped from ports is primarily ore, while the volume of other cargo is 0.17 thousand tons (less than 0.1%).

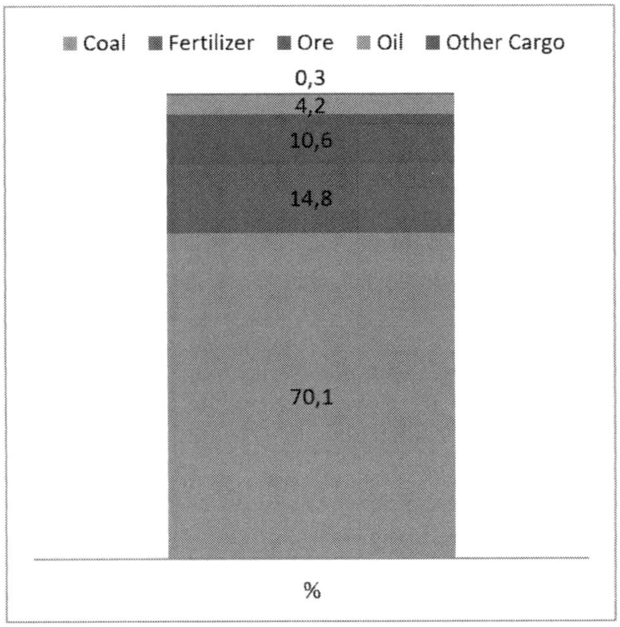

Figure 1. Structure of cargo turnover of the Arctic Basin Ports (cargo shipped to ports), thousand tons (compiled based on the data available at https://cargo-report.info/stat/ports-arctic/)

The current environmental conditions favor the development of the Arctic, while advancing innovative technologies can make resources extraction in the Arctic economically feasible. Most of the resources processed in the Arctic are associated with industries that are harmful to the environment. Developing marine communications also carries environmental risks for the Arctic ecosystem. At the same time, since either industrial corporations or cargo companies are not particularly interested in preserving the natural resources of the region, this task should be undertaken by the state. Sustainable development of the Arctic is impossible without state mechanisms for coordinating the competing interests of various participants. Developing a comprehensive system of administrative and economic measures to compensate for the negative external impacts of the Arctic development is one of the priority goals for the state.

The first step towards this goal may be establishing (at the level of the country, preferably, at the level of the world community) a unique legal framework regulating all types of economic activities, taking into account both natural and climatic characteristics of the region and the specifics of indigenous peoples' lifestyle. Since the development of the Arctic is inevitable and necessary, it is expedient to work out not only the permissible standards of environmental pollution specific for the Arctic, but also the optimal (i.e., permissible at this stage of economic development) pollution levels. Effort should be made to compensate the marginal social costs from the deterioration of the environmental conditions in the region by the marginal social benefits from developing the Arctic territories. Nowadays various methods exist to assess environmental damage from industrial production. However, there are no uniform methods for assessing the social costs from the rising levels of pollution, as well as the social benefits from developing production or transport in the Arctic. Constructing such methods would make it possible for the state to determine the feasibility of the chosen directions of development and to form economically justified mechanisms to compensate negative external impacts. This could include, for example, a system of administrative and economic measures aimed at encouraging enterprises to adopt innovative energy-saving and green technologies.

The system of economic and administrative measures cannot be created without restructuring the institution of environmental impact assessment in the Russian Federation, including:

- increasing the number of objects for assessment;
- establishing the types of activities that cannot be carried out without environmental assessment;
- generating unified requirements to the procedure for conducting the assessment;
- developing a unified procedure for assessing the extent of environmental damage;
- ensuring that the methods and results of the environmental assessment are transparent to the public;
- forming a unified assessment procedure based on the results of environmental impact assessment of external and social costs of intensified economic activities in the Arctic;
- forming a unified procedure for assessing the social benefits from intensified economic activity;
- developing a unified procedure for compensation for the damage caused to the Arctic ecosystems, in the event that a decision is made to either maintain the existing level of economic activity or intensify the activity;
- conducting environmental monitoring in the course of economic activity in the Arctic;
- ensuring transparency of the results of environmental monitoring and developing administrative mechanisms of swift responds to environmental standards violation by suspending the economic activity of the culprits;
- developing a mechanism for prohibiting activity in the Arctic region to enterprises with gross or systematic violations of the environmental standards.

The proposed system of measures should be complemented by additional measures granting economic and other preferences to the enterprises that comply with higher environmental standards, increasing their competitiveness in the development of Arctic resources.

4. Conclusions and discussion

To summarize, the exclusive status of the Arctic is not established in any international documents; for this reason, the legislation supporting environmental monitoring in the Arctic as a whole should be worked out taking into account the specifics of national regulation of environmental issues. Since adopting common international standards is an incredibly complex process, its first stages should involve the coordination of the main provisions of environmental monitoring in the zones of national jurisdiction of the countries bordering the Arctic by at least eliminating the inconsistencies between the legislations. The analysis of legal framework of the Russian Federation that we have carried out revealed that the legislation is not yet ready for developing an environmental monitoring system. The proposed approach can be used to determine the economically feasible level of development of the environmental monitoring system.

Acknowledgment

The study was carried out within the framework of Project no. 0226-2018-0004 of the Institute for Economic Studies "Interaction of Global, National and Regional Factors in the Economic Development of the North and the Arctic Zone of the Russian Federation" as assigned by the state task of IES KSC RAS.

References

[1] Lukin Yu F 2013 *The Russian Arctic in a Changing World* (Arkhangelsk: IPF SAFU Press) p 281 (in Russian)
[2] Voitolovskiy G K Г К 2010 *Bulletin of the Murmansk State Technical University* B **13** 90 (in Russian)
[3] Atland K 2014 *Comparative Strategy* B **33** (2) 145
[4] Peimani H 2013 *Energy security and geopolitics in the Arctic: Challenges and opportunities in*

the 21st Century (World Scientific) p 320
[5] Agarcov S A et al 2018 *Zapiski Gornogo Instituta* B **230** 209 (in Russian)
[6] Korovkin N et al 2017 Ice-Melting process modeling on transmission lines 2017 International Conference on Industrial Engineering, Applications and Manufacturing ICIEAM 2017 - Proceedings
[7] Balashova E S and Gromova E A 2017 Arctic shelf development as a driver of the progress of the Russian energy system MATEC Web of Conferences Available from: https://www.scopus.com/inward/record.uri?eid=2-s2.0-85020221929&doi=10.1051%2fmatecconf%2f201710606008&partnerID=40&md5=592995b6ce6565dd7698a78b788f5fe8 [Accessed 28th May 2018]
[8] Bertelsen R G and Gallucci V 2016 *Marine Policy*, B **72** 240
[9] Blunden M 2012 *International affairs*, B **88** (1) 115
[10] Lalonde S and Lasserre F 2013 *Ocean Development & International Law* B **44** (1) 28
[11] Käpylä J and Mikkola H 2013 The *Finnish Institute of International Affairs* 1-9
[12] Martiynov A S 2007 *Estimation of anthropogenic impacts of large industrial enterprises on the main types of ecosystems in the Russian Arctic taking into account their assimilation potential* Report on scientific reseach. Theme number 1903-08-07 Available from: https://arcticregion.ru/index.php/ekologiya-arktiki [Accessed 20th May 2018]
[13] VanderBerg J D 2018 *Polar Geography* B 41(1) 55
[14] Ng A K Y et al 2018 *Wiley Interdisciplinary Reviews: Climate Change* B **9** (2) 507. doi.org/10.1002/wcc.507
[15] Kaiser B A et al 2016 *Polar Journal* B **6** (1) 152
[16] Tervo-Kankare et al 2018 *Tourism Geographies* B **20** (2) 202

Environmental safety as an element of single-industry towns' sustainable development in the Arctic region

S Gutman, A Teslya

Peter the Great St. Petersburg Polytechnic University, St. Petersburg, Russia

anntes@list.ru

Abstract. Maintaining a certain level of environmental safety is one of the most important goals in exploring the Arctic region, taking into account its intensified industrial development. This study offers an approach to achieving environmental safety of industrial single-industry towns of the Arctic region of Russia. In order to implement an environmental safety strategy insingle-industry towns (with the Murmansk Region discussed as an example) and tailor it to the changing conditions, we have proposed a system of indicators reflecting the goals of preserving and restoring the ecosystem of single-industry towns of the Murmansk Region, taking into account the severe conditions and high vulnerability of the Arctic nature. The indicators were determined using the Balanced Scorecard system which components were adapted in accordance with the goals and objectives of the Strategy of Ecological Security of the Russian Federation in view of the unique features that Arctic single-industry towns have. When choosing the methodology of the study, we have found and analyzed the indicators of sustainable development, including environmental safety, proposed by international communities or organizations, such as UN Commission, World Bank, OECD European Commission, World Wildlife Fund, etc. The strategic map we have constructed allows to reconcile the interests of different stakeholders and can be used as a basis for decision-making at various levels of governance (industrial enterprises, regional and municipal authorities, local communities, and others).

1. Introduction

The issue in creating and implementing programs for industrial development of territories is in reaching certain physical limits of an ecosystem's capacity for self-sustaining. This problem is particularly urgent for the Arctic ecosystem, which is very vulnerable and has little capacity for self-healing. The goal in this region is in reconciling the conflict between economic growth and maintaining a safe environment. It is no accident that international organizations pay great attention to environmental issues, as well as to methods of organizing economic activities aimed at improving people's well-being while significantly reducing environmental risks through using innovative technologies, modernization and increasing production efficiency. This is why analysis of the problems of achieving environmental safety of single-industry towns (SITs) in the Arctic region is an important topic for research.

2. Environmental safety and sustainable development of territories

Both Russian and foreign authors define sustainable development as the balance between economic well-being, social equality and healthy environment; achieving this balance is one of the most important tasks of sustainability [1, 2].

The concept of sustainable development was first formulated in the Brundtland Report (also known as Our Common Future) for the World Commission on Environment and Development, published in 1987 [3]. The report defines sustainable development as the development that meets the needs of the present not compromising the ability of future generations to meet their own needs [4].

The program of sustainable development in the 21st century until year 2030 is described in the Resolution adopted by the UN General Assembly on September 25, 2015 (Transforming our world: the 2030 Agenda for Sustainable development) [5]. According to this document, sustainable development should be based on green economy, which is a complex construction that includes improving energy efficiency via renewable energy sources (RES). The 2009 Human Development Report of the Russian Federation mentioned improving energy efficiency as one of the most important trends of sustainable development [6,7]. Improving energy efficiency could not only reduce greenhouse gas emissions and energy costs but also slow down the depletion of mineral resources and prevent various environmental threats associated with the extraction of these resources [8]. The UN Resolution states that the goals and objectives of the world's sustainable development are a global challenge, universally applicable for all countries. The 17 sustainable development goals and 169 targets are the proof of the new agenda's scale and relevance, continuing the effort already made to protect and restore terrestrial ecosystems, ensure rational use of water resources, provide universal access to reliable and modern energy sources, promote sustainable infrastructure, industrialization and innovation, and so on [6, 9].

Achieving the goals of sustainable development entails strengthening the global partnership. Sustainable development goals are regarded as an additional tool together with international law, treaties and conventions, stimulating the interaction of countries and mobilizing stakeholders to take action [10,11]. The UN has established six criteria of successful sustainable development [12]:

1. People. Ensure healthy lives, promote well-being for everybody at all ages, ensure quality education and inclusion for women and children.
2. Decent living standards. Eradicate extreme poverty and reduce inequality.
3. Prosperity. Promote sustained, inclusive and transparent economic growth.
4. Planet. Protect, restore and promote sustainable use of ecosystems following the interests of all countries and the future generations.
5. Partnerships. Promote global solidarity in the interests of sustainable development.
6. Justice. Promote peaceful and inclusive societies and build effective, accountable and inclusive institutions at all levels.

To quantify the achievement of these goals, the 1992 United Nations Conference on Environment and Development in 1992 raised the issue of indicators for sustainable development [5,13,14]. The original list included 134 indicators; calculation of these indicators complicated the work and analysis at the national level. Later, in 2001, core indicators were determined within the "theme/sub-theme/indicator" framework. The Economic category included two main themes: the economic structure and the production/ consumption patterns. The first consisted of three sub-themes: economy, finance and trade. The second consisted of four sub-themes: consumption of material resources and energy, transport, education, waste management.

OECD's (Organization for Economic Cooperation and Development) system of environmental indicators is another well-known set of indicators. The OECD member countries have been working on a program of environmental indicators since 1990, adhering to special selection criteria: 1. Relevance. 2. Scope of application. 3. Availability for regular use in statistical analysis. The system is based on the "pressure-state-response" (PSR) framework, describing the causal correlation between economic, social and environmental components, which is used to formulate the policy to solve real problems. The OECD indicator system was used as a basis for the system of indicators developed by the World Bank in cooperation with the United Nations Environment Programme and the International Center for Tropical Agriculture in order to improve environmental management. Its distinctive feature is that the indicators are presented in the form of a geographic information system, which facilitates planning and decision-making.

Constructing indicators for sustainable development has been the subject of numerous studies recently. One of the new systems is the SGD Index for monitoring the progress on Sustainable Development Goals, developed by experts from Bertelsmann Stiftung and the UN Sustainable Development Solutions Network (SDSN) at the national level, based on different indices for each specific goal [15].

Various rankings are used in addition to indices and indicators for assessing sustainable development, for example, the regional energy efficiency ranking composed by the Ministry of Energy of the Russian Federation, the ranking for assessment of performance of regional executive authorities composed by the Government of the Russian Federation, regional rankings on fundamental and environmental performance of business (Interfax-ERA), Ecological Footprint of the Russian Regions (WWF), and others. According to the Paris Agreement, due to differing conditions in different countries, absolute emission reduction targets across the economy serve as the key sustainable development indicators in developed countries, while limited or reduced emissions across the economy serve as such in developing countries [16].

According to [17], the global community should be guided by a common strategy while achieving global goals of sustainable development. Neither possession of wealth, nor over-consumption of material goods, but rather indicators of personal merit and social significance should be the key priorities for the future generations; this consideration should be the focal point of such strategies. This will entail changes in political systems, international relations and state mechanisms. It is noted in [18] that regional stability is characterized by a system of indicators, because any given region within the country is a natural and socio-economic systems. The general stability of the region acts as a function of economic, social and environmental sustainability of the region within this approach.

Even though much research has been dedicated to achieving sustainable development [19] and there is general consensus among scholars that consumption of natural goods should not exceed the natural limitations engendered by the parameters of the natural environment of the planet, to date, the value aspects of sustainable development concept have not been studied in sufficient detail, and the mechanisms of achieving the goal have not been determined. The issues of regional differentiation of the sustainable development process (that are especially important for analysis of economic performance in such a unique region as the Arctic) have not been fully considered either.

As evident from the agenda detailed in Our Common Future report of the Brundtland Commission discussed at the 42nd session of the UN General Assembly in April 1987, there has been an emerging understanding that the ecology and economy are integrated in the real world, and that practical economic decisions are made without paying attention to natural resources and other environmental limitations [20]. Even though sustainable development issues are a constant focus of the UN's attention and the organization consistently promotes comprehensive proposals aimed at achieving sustainable development, the principles of sustainable development voiced by the UN are not supported from a theoretical or scientific standpoint or complemented by scientific technologies, methods and systems [21]. The entirety of the links between various natural structures, social institutions and human activity structures should be thoroughly studied, understood and described in order to make a comprehensive transition to sustainable development in the ecology-society-economy system [22]. Sustainable development can be achieved through developing the scientific method, based on the general laws within the nature-society-man system compiling natural, social and spiritual processes.

Let us assess the readiness of the society for interaction on environmental issues, with Russia as an example. The term "green economy" was first introduced about 30 years ago by British economists in their study "Blueprint for a green economy" [23]. In 2010, UNEP produced a general definition of this concept [24]: the green economy can be defined as one that results in improved human well-being and social equity, while significantly reducing environmental risks and ecological scarcities. Russian Federation has been implementing a "green" policy based on the Constitution, the principles and norms of international law, international treaties signed by the Russian Federation, and on additional laws, orders and decrees issued by the state authorities [25]. In recent years, the Government of the Russian Federation has set many goals based on the principles of the green economy: introducing eco-innovations and improving sustainable energy development in Russia, among others [26]. The Federal Laws aimed at strengthening the system of environmental monitoring standards and carrying out measures to stimulate the Russian economy were amended in late 2017 [27, 28]. Decree of the President of the Russian Federation No. 176 of April 19, 2017 set out the Strategy for Ecological Security of the Russian Federation up to 2025, establishing the goals of the state environmental policy: preserving and restoring the natural environment, ensuring the quality of the environment necessary for human well-being and sustainable economic development, eliminating the accumulated environmental damage due to intensified economic and industrial activity as well as global climate change. The Strategy also established the following main targets, priorities and indicators for environmental impact assessment [29].

Complying with the principles of green economy, such as energy efficiency and energy conservation (in the energy sector), environmental sustainability of agriculture, "green" construction, "green finance", etc. contributes to sustained and sustainable development of territories. [30].

The draft of the National Public Standard "Arctic Region: Environmental Safety" recommends that, in order to ensure sustained and sustainable development of the Arctic, each region of the Russian Federation develops, approves and implements a socio-economic development strategy, which must include a section on developing the infrastructure for environmental management of the region. The strategy should describe the target programs and separate projects and measures sharing common goals, objectives, schedules and resources, ensuring the development of the environmental safety system in the region of the Russian Federation. The action plan for ensuring the environmental safety of the region (roadmap) can be adopted either as a separate document or as a section of a more general document describing the strategy for regional development, and should be approved in the manner set out in the general document. A separate section of this document should be dedicated to motivating enterprises to implement standardization policies [31].

Sustainable development is the only possible path for the Arctic region due to a number of geopolitical and geo-economic characteristics, as well as to the specific goals established for the region. The effort being made to organize a continuous and timely supply of resources, on the one hand, and the scope of the global interconnected problems arising during the technological development of the Arctic, on the other hand, are the reasons behind the world community's commitment towards ensuring environmental safety during the development of the territories of the Russian Arctic.

Environmentally safe technologies must be adopted in developing the Arctic territories due to the following factors:

- growing interest of the world community in the problems of environmental safety in the Arctic and rapid development of international environmental law;
- increasing demand for environmentally friendly technologies actively used by other countries;
- depletion of traditional energy resources increasing their costs;
- the need for reducing the risks of climate change;
- the need for reducing the risks of large-scale man-made disasters;

At the regional level, the environmental safety of the Arctic territories is governed by the existing techno-social and natural system. Industrial production is a source of harmful factors determining the territorial indicators associated with the actual and standard levels of man-made impact. [32] Therefore, from the standpoint of achieving environmental safety, man-made impact of the regional industry acts an "internal" hazard. The following steps should be taken to assess and develop a set of measures ensuring the environmental safety of the territories with the goal of achieving sustainable development of the region as a territorial techno-social and natural system:

1. Carrying out comprehensive analysis of regional environmental safety;
2. Assessment of the environmental safety level investigating the main sources of man-made impact, as well as the main enterprises posing risks of negative impact.

3. Carrying out a forecast of the changes in the main parameters of man-made impact on environmental safety, taking into account the planned changes in the structure of industrial production and territorial characteristics of the region that are unavoidable in the integrated development of the Arctic.

4. Confirmation that the obtained data are used when implementing the regional development strategy (including tax, credit and financing policies, social policies, etc.).

The above-given factors pose the problem of selecting a system of indicators for comprehensive assessment of the region's environmental safety, for coordinating different aspects of regional governance, as well as for reconciling the conflicting interests of different stakeholders (industrial enterprises, regional and municipal authorities, the local community, and others). In our opinion, the balanced scorecard system (BSC) should serve as a methodological basis for solving this problem. The strategic maps developed by the cascading method (balanced scorecards for individual levels of management, such as regional and municipal levels) make it possible to coordinate the competing interests in devising and implementing a territories' sustainable development strategy. We generated a BSC for each level of the organizational unit keeping in mind that, unlike commercial companies, the authorities have to focus on the stakeholder (or customer in the classical version of the BSC) component, instead of the financial one [33].

3. Single-industry towns and the problem of environmental safety control

The problem of maintaining environmental safety along with sustainable development of territories is especially important for Arctic SITs, where the town-forming enterprises are mainly those of the mining industry. We have used the approach described in [34], with the general stability of the Arctic region regarded as a derivative of certain activities in the region (economic, social, environmental). This paper, which is a part of a more general study, focuses on achieving environmental safety of SITs while ensuring sustainable development of the region.

The socio-economic situation in Arctic SITs is particularly severe, with the population steadily declining. A set of measures aimed at achieving a strategy for maintaining the environmental safety of SITs should take into account the specifics of the development of the territories, including a strong direct dependence of all the factors affecting the population's quality of life on the performance of the town-forming enterprise.

The solution traditionally proposed for ensuring sustainable development of SITs is in reducing the SIT's dependence on the town-forming enterprise, diversifying economic activity and developing small business. Not denying the need for infrastructural development and diversification, we consider it equally important to reconcile the interests of the three main participants in the economic activity of a SIT, i.e., the municipal government, the town-forming enterprise and the local community, and to form a sustainable, mutually beneficial partnership. This is facilitated by:

- relative geographical isolation of SITs of the Arctic region which means that the interests of the inhabitants are more or less similar, and the variance of their needs is low.
- the town-forming enterprise and municipal authorities heavily depend on each other, making interaction especially important.
- the number of subjects with competing interests is relatively small, as is the number of priority goals.

Achieving environmental safety is a special activity sphere that has all the conditions and opportunities for reconciling the interests of stakeholders, since there is a sufficiently well-constructed and evolving system of laws regulating the environmental activities. it is possible to obtain a quantitative environmental assessment of industrial emissions; there are scientific methods offering sets of indicators for comprehensive analysis of environmental issues in the region, while regional and federal authorities are concerned about this problem. This means that reconciling the interests in the environmental sphere should be easier than, for example, in solving the problems of the social sphere. Effective partnerships should be based on balancing the interests of stakeholders taking into account the demands of each participant (the municipal government, the town-forming enterprise and the local community) in matters of environmental safety.

Analysis of business activity in SITs of the Russian Arctic has revealed that small business is rather poorly developed, unstructured and cannot be regarded as a key stakeholder at present [35]. However, in generating a map for achieving environmental safety, we have taken into account the issues of providing support for small business, making it possible to include small business as the fourth element of the system of participants.

4. Environmental safety map (with single-industry towns of the Murmansk Region of the Russian Federation as an example)

The Murmansk Region was chosen as the study object because of its high percentage of SITs situated in the Arctic region. Arctic SITs of the Murmansk Region also have common traits and development issues, and their town-forming enterprises are mostly those of the mining industry. The common problems shared across the region allowed to form a strategic map for ensuring environmental safety with the goal of achieving sustainable development of SITs of the Murmansk region (see Fig. 1). The goal is to ensure the required level of environmental safety of SITs of the Murmansk Region, i.e., such a level of the system's development that satisfies the following conditions:

- enforces the currently existing environmental standards for the main sources of man-made impact;
- corresponds to the goals and objectives determined by the current regional strategy of sustainable development;
- ensures that the long-term goals and objectives are achieved in accordance with the regional strategy of sustainable development for the future.

Thus, the issue is not only and not so much in preserving the region's resources, but in improving the well-being of its population, with the environmental component as a contributing factor. This is the reason why improving the well-being of the stakeholders is at the top level of the strategic map.

The map is based on the following priority goals important for SITs of the Murmansk Region:

- preserving and improving the labor resources of the SIT;
- using innovative environmentally efficient technologies;
- developing the small business sector and increasing the stability of the SIT's economy;
- searching for funding sources for environmental safety projects and substantiating their attractiveness.

"Finance and budget" is the component acting as a source of development and as the basic level of the strategic map. We have deliberately elected not to use the traditional framework prescribed by the methodology, since searching for sources of funding for environmental projects is the most urgent task for SITs, where the severe socio-economic situation prevents the towns from becoming the sources of financing for the projects. On the other hand, as the town-forming enterprises are participants of larger industrial associations, they do not fully control their financial resources. In this case, the state has to take steps to support environmental monitoring (i.e., compliance with norms and standards) and to create the most favorable conditions motivating the town-forming enterprises to invest in green technologies reducing the anthropogenic pressure on the environment.

We should note that introducing innovative technologies can have a high multiplier effect, providing employment for the population, increasing business activity in other sectors of the economy, stimulating innovation, making it possible to effectively develop resources and territories even given that the infrastructure of the Arctic SITs is rather underdeveloped. Thus, achieving environmental safety should not be just a cost-based project, but on the contrary, combine the goals of environmental protection with boosting the competitiveness of the town economy, improving the quality of life of the population, achieving both environmental and economic security.

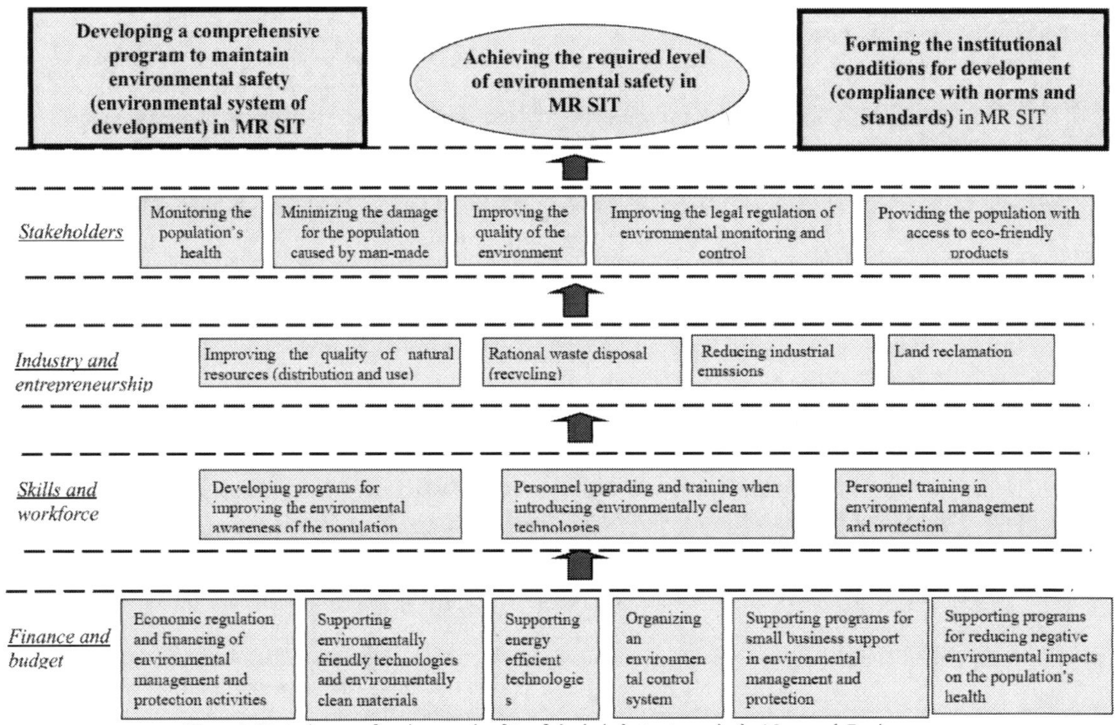

Figure1. Strategic map of enviromental safety of single-industry towns in the Murmansk Region
Abbreviations: SIT stands for single-industry towns, MR stands for Murmansk Region

5. Directions for future research and discussion

This study presents a strategic map for reconciling the interests in achieving the required level of environmental safety for the Murmansk Region of the Arctic zone of the Russian Federation. While the map discussed in this study was constructed taking into account the specifics of the Murmansk Region as a whole, similar maps should be developed in the future for each of the region's single-industry towns, taking into account the specifics of these towns, their priority goals and objectives, and the extent to which the town-forming enterprises are ready for cooperation.

A separate issue is choosing the specific indicators characterizing the goals and problems in achieving sustainable development, taking into account the tasks of preserving and restoring the Arctic ecosystem in view of its severe environmental conditions and high vulnerability.

The National Public Standard "Arctic Region: Environmental Safety" that is currently under development combines the best international and Russian practices in the field of public participation in environmental control and can serve as a basis for selecting such indicators. However, the standard has to be adapted as it does not take into account the specifics of single-industry towns.

References

[1] Uskova T V 2009 *Management of sustainable development of the region* (Vologda (ISEDT RAS) p 355 (in Russian)

[2] Gibson E 2014 Supporting sustainable technology cluster development: A performance measurement problem (Conference Paper) *Portland Intern. Conf. on Management of Engineering and Technology*

[3] *Our common future* Report of the International Commission on Environment and Development 1989 (Moscow (Progress) p 21 (in Russian)

[4] Kasimov N S et al 2004 *Bulletin of the Russian Academy of Sciences* B **74** (11) 28 (in Russian)

[5] Resolution by the UN General Assembly on September 25 09 2015 *Transformation of Our World: An Agenda for Sustainable Development for the period up to year 2030* Available from: http://www.un.org [Accessed 18th Januaruary 2018] (in Russian)

[6] The future that we want. A summarizing document of the F 2012 United Nations Conference on Sustainable Development Rio de Janeiro Brazil p 20 (in Russian)

[7] Bobylev S N et al 2010 *Report on Human Development in the Russian Federation 2009. Energy and Sustainable Development* (Moscow (UNDP) p 270 (in Russian)

[8] Bobylev S N et al 2013 *Report on Human Development of the Russian Federation 2013. Sustainable development: challenges of RIO* (Moscow (OOO «RA ILF ») p 202 (in Russian)

[9] Bobylev S N et al 2015 *Report on Human Development of the Russian Federation 2015. Human development in a downturn economy* (Moscow (Analytical Center under the Government of the Russian Federation) p 260 (in Russian)

[10] Weber A 1998 *Free thought* B **5** 42 (in Russian)

[11] Framing Sustainable Development Goals, Targets, and Indicators PREPARED BY THE SDSN SECRETARIAT 2014

[12] Universal Sustainable Development Goals. Understanding the Transformational Challenge for Developed Countries.Report of a Studyby Stakeholder Forum 2015 p 25

[13] Indicators for Sustainable Development: guide lines and methodologies. Available from: https://sustainable development.un.org [Accessed 2th February 2018]

[14] Indicators for the Integration of Environmental Concerns to Transport Policies. Working Group on the State of the Environment. OECD 1999.

[15] SDG Index & Dashboards. A global report (full version) Available from: http://www.sdgindex.org/ [Accessed 08th June 2018]

[16] Framework Convention on Climate Change 2015 UN Adoption of the Paris Agreement. Chairman's proposal The solution draft CP.21. p 42 (in Russian)

[17] Barlybaeva X A 2014 *Sustainable development. Globalistics* (Moscow (Publishing house

"Scientific Library") p 39 (in Russian)
[18] Tsapieva O K 2010 *Problems of modern economy* B **2** 307 (in Russian)
[19] Leksin V N and Profiryev B N 2017 *Economy of Region* B **13**(4) 985 DOI: 10.17059/2017-4-2 (in Russian)
[20] Wyckoff A Colecchia A 2011 The Economic and Social Impact of Electronic Commerce (Organization for Economic Co-Operation and Development (OECD),. P.11.
[21] Neudakhina Yu G 2011 *Main directions of research of sustainable development problems in the region.* Dissertation of FSU HPE "Kabardino-Balkar State Agricultural Academy". In the name of V. M. Kokov. Ed. Nalchik.P. 233.**15**(1) 450 (in Russian)
[22] Zalikhanov M *et al* 2002 The scientific basis of the sustainable development strategy of the Russian Federation (Moscow (Publishing house of the State Duma) p 207 (in Russian)
[23] Pearce D W 1989 *Blueprint for a green economy: economic report* (London: (Earthscan Publications Ltd) p 192
[24] UNEP «Green Economy: developing countries success stories» 2010 ЮНЕП Kenya Available from: www.minambiente.it/ [Accessed 05th May 2018]
[25] OECD green growth indicators 2017. Available from: https://read.oecd-ilibrary.org/environment/green-growth-indicators-2017_9789264268586-en#page1 [Accessed 20th June 2018] (in Russian)
[26] Analytical Herald No. 11 (668). Council of Federation (04/26/2017) COUNCIL.GOV.RU official site of the Federal Assembly of the Russian Federation Available from: http://www.council.gov.ru [Accessed 14th May 2018] (in Russian)
[27] Federal Law No. 458-FZ of December 29, 2014 "On Amendments to the Federal Law" On Production and Consumption Wastes" Available from: http://kremlin.ru/acts/bank/39278 [Accessed 16th May 2018] (in Russian)
[28] Federal Law of 21.07.2014 No. 219-FZ "On Amendments to the Federal Law" On Environmental Protection "and certain legislative acts of the Russian Federation Available from: http://kremlin.ru/acts/bank/38787 [Accessed 16th May 2018] (in Russian)
[29] The Strategy of Ecological Safety of the Russian Federation for the Period Until 2025 Available from: http://kremlin.ru/acts/bank/41879 [Accessed 24th May 2018]
[30] Kudinova G et al 2012 *Interdisciplinary scientific and applied journal "Biosphere"* B **4**(3) 245 (in Russian)
[31] National public standard "Arctic environmental safety" 2016 (Moscow Publishing house "Libri Plus") p 88 (in Russian)
[32] Didenko N, Rudenko D and Skripnuk D 2015 Environmental security issues in the Russian Arctic *Conf. Proc. of Int. Mult. Sc. GeoConf. Surveying Geology and Mining Ecology Management, SGEM* **3** (5) 267-274
[33] Gutman S S and Teslya A B 2016 *Murmansk State Technical University Vestnik* B **19** (2) 383 (in Russian)
[34] Bell S and Morse S 2012 *Sustainability indicators: measuring the immeasurable?* (Cornwal (Typeset by MapSet Ltd Gateshead Printed and bound in the UK by TJ International Ltd Padstow) p 251
[35] Gutman S S et al 2015 Restructuring of the economy and engineering education: problems and development prospects *Coll. of Works of the Scientific and Practical Conf. with International Participation* 306 (in Russian)

Synergies of the Economic Development of the Arctic Cluster System

E A Afonichkina[1], **A I Afonichkin**[2]

[1]Peter the Great St. Petersburg Polytechnic University, St.Petersburg, Russia
[2]Samara National Research University, Russia

ekaterinaafonichkina@gmail.com

Abstract. The paper discusses the problem of modelling the development policy of the Arctic Cluster of the Russian Federation. This economic development task takes into account the complex conditions and necessary requirements for the coordinated growth of all participants in the Arctic Cluster in the form of complex, spatially sectorial economic systems, including international. An important condition for such a policy is the possibility of generating synergy in the development process, from joint participation in the implementation of the development programme, especially in the face of significant constraints and balanced requirements. Failure to meet the conditions for the formation of synergy generates development risks. A network model of the Arctic Cluster is proposed, whose formula, describing the generalised strategy of system-wide development, is presented in the paper. Definitions of the formation of synergy of stage-by-stage development are also given. A formula combining the results of the development of the synergy effect at each stage to show cumulative systematic synergy is presented.

1. Introduction

One of the important strategic zones, which contains potentially broad opportunities for economic development, but also has very significant systematic risks, is the Arctic management zone. The effective use of the Arctic zone has long been a topic of consideration; however, the conditions and requirements for operating, managing and developing this segment are very high and require careful economic analysis. The attractiveness of this zone is that it disposes significant volumes of valuable natural resources and a high potential for the economic development of circumpolar countries. However, the level of risks is so great that full development of this zone is impossible without joint strategic activities of the countries concerned. In particular, the following Arctic segments are of economic interest: the waters of the Arctic ocean, the continental (land) territory of the zone and the continental shelf of the Arctic Territory. It is here that reserves of oil and gas, coal, gold and other valuable metals are concentrated. Opportunities for the formation of efficient transport and transit systems along the Northern Sea Route are also a topic of great current interest [1,2].

Despite the fact that almost all of the circumpolar countries of the Arctic zone have Arctic development strategies, their effectiveness depends both on the volume of investments as well as on a number of special conditions. The development of the Arctic zone is limited by the following factors [1,3,4]:

- Extreme climate, presenting particular management challenges;
- Weak resilience of the Arctic ecology to external and anthropogenic influences;
- Underdevelopment of transport and logistics as well as production and service networks;
- Spatial disunity of regions;
- Current lack of Arctic-specific life support technologies;
- Economic and legal inconsistency of participating countries.

Overcoming unfavourable conditions is only possible with the concerted interaction of the parties concerned in the implementation of Arctic projects. This requires both political will and a balancing of economic interests in the development of the global Arctic zone. In particular, the industrial development of the Arctic is written into Russia's long-term development strategy [1,2]. Its effective development requires consideration of a number of specific spatial and sectorial factors, the formation of interconnected international industry and interindustry corporations, complementary industries and expansion of foreign economic activity, at least in this zone. However, the solution of such problems is connected to existing difficulties of the scientific and technical, economic, socio-environmental as well as international and legal aspects of the development of the Arctic zone.

The system of interactions of inter-branch economic systems within the spatial cluster of the Arctic leads to the formation of a specific Arctic spatial-sectorial economic system, in which not only interindustry systems, but also international alliances are involved. Typically, vast undeveloped spaces have considerable synergetic development potential; in this case, however, this would require cooperative adaptation of the entire Arctic infrastructure. All this necessitates the further expansion and deepening of research on problems in the development of economic activity in the Arctic zone for the effective management of changes in this economic space, taking into account the interests of all participating countries.

2. Network Model of the Arctic Cluster

The possible formation of a single Arctic economic zone in the form of a structured international cluster system can use the priorities of economic systems integrated into the spatial (Arctic) strategic economic zone in its development taking into account the interests of the participating circumpolar countries. Such systems allow the rational and purposeful management of processes of structural formation and development of the Arctic Cluster economy, its spatial organisation and the coordination of growth with the goals of sectoral development strategies, contributing to the creation and development of large territorial and sectorial economic complexes that serve as drivers for the economic growth of the Arctic Territories.

The affiliation of such systems to different international jurisdictions, while pursuing common development interests, has the potential to yield an effective result having the essential component of synergy. The international integration of the Arctic zone as a strategic economic zone for international and inter-industry corporations provides integration incentives leading to the formation of an economic system whose primary structure comprises a cluster network. Such a cluster network (*CN*) comprises an international entrepreneurship system within which economic entities having coinciding economic interests can enter into coordinated economic and managerial interactions on a contractual basis. Undoubtedly, in this case, it is necessary to clearly define the areas of responsibility, interests and goals of management, development resources, principles for coordinating the interests of cluster network participants in order to increase competitiveness and maximise synergies [5,6]. However, in such complex systems, which have strong economic convergence, a "backward wave" effect may appear, leading to a stronger and uneven concentration and zoning of the economic space, increasing not the synergy and development risks. Therefore, the policy of economic development of such cluster networks should be balanced and take into account the possibility of spatial and economic differentiation and risk alongside the pursuit of the multiplicative effect.

In many respects, development policy depends on the model of the system itself: for objects such as the economic cluster, the basic policy will be somewhat simplified, failing to reflect the complexity and effectiveness of cluster network agent interactions and the conditions for their implementation in the Arctic zone in international and inter-industry interactions. On the one hand, such integration of spatial and sectorial economic systems provides significant opportunities in supporting development processes due to synergies of integration, mutual influence of resources and opportunities, strengthening of investment attractiveness, growth of competitive advantages, etc. On the other hand, such integration generates a significant number of systemic risks associated with the generalisation of the management policy and the formation of a unified development strategy that takes into account the interests of all participants in this spatially sectorial economic system. The conditions for the formation of the synergistic effect of cluster development can be described as necessary and sufficient conditions for balancing and cohering activities of participants in the economic relations of the cluster network, as well as individual technological chains of the entire economic system of the Arctic zone.

In accordance with the typologisation of clusters, let us supplement and structure the cluster network model with the following elements:

- Interacting elements of technological chains (*TC*) and, as per cluster network, participants in the technological chain, the total number of which comprises accounts K, $U_{TC}=\{u_1,u_2,u_3,\ldots,u_K\}$, $k=1–K$.
- Participant (an active element, a^0) of the cluster network (*CN*), defined as the cluster management centre (of different possible levels of management). Let this comprise index $k=0$. Such a participant performs control and coordination functions for the remaining elements of a separate segment of the network CN_j and of the entire cluster network as a whole $\{CN\}$.
- Set of operations that require interactions ($\{a_i\}\{R_{CN}\}a_j$) with other cluster members (a_i), the number of which for the jth participant equals to N_k, $\{R_{Nk}\}$. Moreover, each type of interaction (R_{CN}) is determined by different types of relationships: – technological (R^T), – informational (R^I), organisational and managerial (R^U), – financial (R^F), etc., $\{R_{CN}\}\subseteq\{R^T,R^I,R^U,R^F,\ldots\}$.

The power (intensity, strength level) of interactions can be different and is determined by a function V_σ^R expressing the intensity dependence for specific types of interactions ($a_i\rightarrow a_j$), or $(a_i)\bigcap_{j=1}^{n}R_j(a_j)$.

External organisations (b_i) (interacting with elements of the cluster system) that are not integrated into the cluster network structure, the number of which is equal to T_j, $\{R_{Tj}\}$, determine the types of interactions; the index j defines the element $a_j\subseteq\{CN\}$ with which the given external interaction takes place – $j=1,2,\ldots T_j$, $\{b_i\}\bigcap_{j=1}^{n}R_{Tj}(a_j)$ or ($\{b_i\}\{R_{Tj}\}a_j$).

Let us determine that the model of the cluster network $\{CN\}$ is determined by the state of the economic potential, denoted as S_{NK} and referring to the value of the control vector U_{TC}, $\{CN\}$, a^0, ($\{a_i\}\{R_{CN}\}a_j$), ($\{b_i\}\{R_{Tj}\}a_j$), $u_k\subseteq U_{TC}$ $u=\{u_k, k=0\ldots.K\}$, $\{u_0,u_1,\ldots u_K\}\subseteq S_{CN}$, where each element $a_j\subseteq\{CN\}$ (within the framework of the *TC*) has its own target control vector z^k and additionally provides the system-wide goals z^O. Thus, the target vector of the element is defined in the form $z_k=z^O+z^k$. We believe that the target vector of the element a_j can be achieved following implementation of a system development strategy aimed at developing both the most active element (*TC* or $\{CN_j\}$) and the system-wide element $\{CN\}$).

We assume that z^k determines the strategy vector for the development of a particular segment of the cluster network $\{CN_j\}$,$\{p_1, p_2,\ldots,p_n\}\subseteq\{CN_j\}$, and z^O is the system-wide development vector $\{p^o_1, p^o_2,\ldots,p^o_k\}$, where the index k characterises part of the system-wide directions within the strategy of each element of the cluster network $\{a_j\}\subseteq\{CN_j\}$.

Then the generalised strategy formulated within the strategy of the elements $\{a_1,a_2,\ldots,a_k\}\subseteq\{CN\}$ for achieving system-wide development goals will be: $u=f(z)$, $U=\sum_{k=1}^{K}u_k+u^o$, $z_k=z^O+z^k$, $z^k=f(\{p_1, p_2,\ldots,p_n\})$, $z^O=f(\{p^o_1, p^o_2,\ldots,p^o_k\})$.

With this, we find that a system-wide development is determined by a generalised strategy of the form:

$$U=\sum_{k=1}^{K}z^k+u^o=\sum_{k=1}^{K}\sum_{n=1}^{N}p_{kn}+\bigcap_{k=0}^{K}p_k^o \qquad (1)$$

Thus, expression (1) establishes the priority directions for the development of the cluster network. The economic goal of the cluster network operation is the maximisation of the vector generalised criterion for the effectiveness of the cluster network, Q, taking into account the synergy of the system development ΔE

$$Q=\{\{a\},u^k,\{R\},U,p_k,p^o,E^k,\Delta E\} \qquad (2)$$

where the factors of the vector are particular criteria for the efficiency of E^k for each active element, technological chain or set of elements of the cluster network.

Thus, for the governing body of the cluster network (a^0), the problem of effective development is reduced to determining the control vector that maximises criterion (2), taking into account:

- Existing restrictions on resources, investment and the initial development potential of each element $\{a_1, a_2, ..., a_k\} \subseteq \{CN\}$;
- The agreed development strategy, i.e. portfolio of cluster network development;
- Coordination of interests and results of activity between the participants in accordance with the rules of balance of interests.

3. Synergetic Effect of the Development Process

The overall economic effect represents the total result of the economic activity of the whole system, taking into account the result of each individual participant. The efficiency of each participant's activity (E_{sj}) entering the *CN* can be represented in the form $E_{sj} = (P/C)$, where P – is the economic result in the process of carrying out the changes; C – is the cost for carrying out the changes. The resource indicator can also be used in the form of the overall profitability of the participating enterprise. If the economic result is presented as the net profit *(Pr)* of the participant for the period and the costs of the changes are defined as the enterprise's investments for the period under study – that is, its assets *(A)* – then the participant's performance can be estimated by the following formula: $E_{Sj} = (Pr / A)_j$.

Conversely, the system synergy is defined as $E_S = \{E_{S1}, E_{S2}, ... E_{Sn}, \Delta E_S\}$, where ΔE_S is the synergistic effect generated by the *CN* in the course of its development. Taking into account the distribution of the *CN* result among the participants in accordance with some rule ρ, we obtain $E^S = E_s + \Delta E_s = \prod_{j=1}^{n} (\rho_j * \frac{\Pr_j}{A_j}) + \Delta E_s$, where n – number of participating companies, $j = 1 - n$.

The synergistic effect – ΔE_S, can be manifested in the corporate development process for *CN* with a large number of structural elements of the system, with individual functions and tasks being implemented jointly. Synergy can be generated by each *CN* participant operating in *CN* value chains in the form of the additive function $\Delta E^E{}_S$ (synergy from mutual participation), $\Delta E^E{}_S = (\Delta^E E_{S1} + \Delta^E E_{S2} + ... + \Delta^E E_{Sn}) = \sum_{i=1}^{n} \Delta^E E_{Si}$. In this case, synergy complements the level of efficiency of the development process through a variation of synergy factors along the development directions under certain conditions and constraints. Denoting ΔE_S as an incremental vector of the additional effect (synergy) in the directions of development, due to systematisation, coordination, economy, balance of resources and development potential in all directions of cluster network development, we obtain: $\Delta E_S = \sum_{k=1}^{m} \sum_{j=1}^{n} \Delta^k E_{Sj}$, where k – type of generated synergy, $k = 1 ... m$, j – corporate network member *(CN)*, $j - 1 ... n$.

Direct implementation of the coordination of such interactions, taking into account the level of synergy and implementation costs, is performed by the cluster management centre (a^0), which has the appropriate rights and powers.

To model the synergy factors in the sustainable development management process, let us outline the structural features of the cluster system as follows:

- Unity of the system goal;
- Joint activities of participants;
- System of interactions of participants in the cluster network;
- Consistency of interests of participants;
- Autonomy of the activities of individual participants;
- Collective and mutual responsibility for the results of joint activities;
- Specialisation and complementarity of functions and operations in value chains and networks;
- Stability of the structure of *CN*;

- Presence of synergy of the development of the *CN*.

In this connection, it is possible to outline the following range of tasks for the effective development of a cluster network with many conditions and limitations:

- Formation of a system of *agents* (*the composition of the CN participants*), distributed in the sectorial and spatial aspects of the economic system;
- Modelling the behaviour of economic agents of the spatial-sectorial economic system (cluster system);
- Modelling of coordinated interactions of *CN* agents.

The description of the dynamics of collective behaviour in the cluster network establishes the level of development dynamic synergy, where the behaviour of each of the network agents (cluster members) at the time t_0 is determined by its own state S^0 in the life cycle stage space, as well as by the states of other participants in the corporate network $\{S^1,S^2,\dots,S^N\}$ at the preceding time ($t_{0\text{-}j}$) $\{S^0,S^1,S^2,\dots,S^N\}/t_{(0+j)}$ and the interactions between them. These factors make it possible to distinguish the following components of the synergy of the cluster development process:

- Synergy of the coordinated interaction ($\Delta Э^S$) (sectorial elements ($\Delta Э_O$), spatial (ΔE_P), spatial-sectorial ($\Delta E_{P\text{-}O}$), provided that $\Delta E^S = \Delta E_O \bigcup \Delta E_P \bigcup \Delta E_{P-O}$;
- Synergy of adequate distribution of system results by participants (ΔE^{SR});
- Synergy of the phased coordinated development (by the stages of the life cycle) (ΔE^{LCS});
- Synergy of system development (multiplication for all previous components) (ΔE^{CN}), as a function: $\Delta E^{CN} = \Delta E^S \bigcup \Delta S^{SR} \bigcup \Delta E^{LCS}$.

From the position of control theory, the model (CN) can be described by the following parameters: $A^j{}_K$ – the set of agents (j = 1 – sectorial, j =2 – spatial, j =3 – sectorial-spatial participants, j = 0 – cluster network control centre); index K – specifies the number of agents of the corresponding type, from which the cluster network is constructed $A^j{}_K = \{(a^1{}_1,a^1{}_2,\dots a^1{}_n),(a^2{}_1,a^2{}_2,\dots,a^2{}_d),\dots,(a^0)\}$.When forming the CN structure, the participating agents form a $\{R_{ij}\}$ a system of interactions to solve operational problems and development problems, generating the E^S -function of the efficiency of the CN, $Ef(N) = F(\{P\},A^j{}_K,\{R_{ij}\})$. The system of relationships $\{R\}$ can be defined in the general case by a matrix of interactions, which can also be asymmetric. $\{R_{ij}\} = R(T_{ij},X_{ij},V_{ij})$, $R_{ij} = f(a_i,a_j,\dots,a_z)$, where: T – the matrix of typologies of interrelations between the elements of the network $(a_i,a_j,\dots,a_z)$; X – the matrix of directions, resources (potential) of interactions and target parameters of interaction between network elements $(a_i,a_j,\dots,a_z)$, determining the link function of the network vertex a_i, with vertex a_j; V – matrix on the levels of interaction strength for the participants in the cluster network. Thus, the vector of interactions R_{ij} is a set of independent matrix parameters characterising the interrelations and interactions between the members of the cluster network $(a_i,a_j,\dots,a_z)$, i.e. those that cannot be further subdivided into constituent components. We will assume that these parameters will be conditional-constant for a certain time interval. Taking into account the multiplicity of relationships, multidimensional interactions $R^k{}_{ij}$ – interconnections of power **k**, such as $R^k{}_{ij} \subseteq \{R_{ij}\}$ can be realised between the participants (a_i,a_j).

If the increment of efficiency due to synergy is determined through the function of synergy F, then E_S can be defined as the efficiency weighted by the effect of the synergistic lever $(1+\Delta)$, $E_S = E + \Delta E = E(1+\Delta) = Ef{*}F$, $F = 1+\Delta$. The set of local synergetic effects for each participant can be divided into synergies of coordination and interaction, synergies of cyclic (stage-by-stage) development, managerial synergies, etc. We will briefly consider some of the different types of development synergy in the next section.

4. Synergy of Coordination and Interaction of Participants

The coherence of interactions in the process of economic development determines the level of balance of interests and input resources of participants in the cluster network. It thus remains a necessary component for generating the maximum corporate development synergy [7,8,9]. Thus, we formulate a number of conditions for assessing the degree of balance and the level of development synergy:

- The leading role of consistency indicators should belong to the financial performance characteristics (profit, revenue, productivity, assets, etc.), which set possible 'growth points' and prospects for development;
- Interactions should describe various areas of activity of the cluster network and the state of participants in the dynamics;
- The level of consistency should be based on the use of a system of temporal and balanced indicators;
- The *SB* should take into account the various components of synergy in the processes, participants and stages of development of the cluster network;
- The synergy effect should provide the maximum system effect.

In this connection, a determination of the criterion of coherence of interests of participants in the cluster network (*CN*) (financial, material, efficient, informational, etc.) and the formation of its maximum value ensures the generation of a marginal synergistic effect. The balance of interests in the process of development management is an important condition for the formation of *CN* participant coherence during the period of growth and development since system efficiency can vary quite significantly as a consequence. In the literature, approaches for an assessment of the degree of balance between corporate interests and areas of activity are described. These tend to primarily feature financial performance indicators (ratings, performance indicators, temporal indicators, etc.) A growth index is used for assessing the current state of corporate governance and determining its level of consistency.

The degree of coherence is assessed through the level of conflict in the management system, for which purpose the index of the intensity of the corporate conflict of interests (including financial) of the *CN* participants:

$$CCI_i = \frac{\sum_{k=1}^{13} d_k w_k}{\sum_{k=1}^{13} w_k},$$

where CCI_i is the index of the intensity of the corporate conflict of the ith participant of cluster interactions, d_k is the variable characterising the presence or absence of the kth attribute, w_k is the weight of the kth attribute.

However, the level of synergetic effect depends not only on the balance of interests, but also on the adequacy and accuracy of the interactions and mutual relations of participants with each other, reflecting the effect of the multicollinearity of the participants themselves, i.e. showing the change in the effectiveness of the participant S_j under the influence of the participant S_i. This can be reflected through the corrective indicator $RE_{Si,Sj}$, which increases or decreases the level of synergetic effect of the participants' interaction. The multiple correction index $RE_{S1,S2,...,Sm}$ can be used to assess the degree of interaction of several participants $\{S_1,S_2,...,S_{m-1}\}$ with the change in the share of the participant's synergy S_m.

Taking into account the additivity of the components of synergetic effects and possible adjustment thereof in terms of the interaction synergy of participants, let us define a generalised expression that specifies a quasi-separable function: $\Delta E_S = \sum_{k=1}^{m} \sum_{j=1}^{n} \Delta^k E_{Sj} + \sum_{k=1}^{m} \Delta^k E_{Sj} \prod_{i=1}^{n} RE^i{}_{S1,S2,...,Sn}$. Each of the corrective indicators RE^i can have both '+' and '-' signs; this depends on the conditions for reconciling interests, resources, growth potential and the level of contribution to the activity and effectiveness of the cluster system.

5. Synergy of Stage-By-Stage (Cyclic) Development

In accordance with the stages of the development life cycle, the synergy of development is determined by group laws, factors and conditions that variously affect the effectiveness of economic systems [10,11,12,13]. Let us formulate some definitions of the formation of synergy in the process of cluster network development.

Synergy of phased development. Analysing the formation process and the factors that generate synergy, we believe that the development synergy manifests itself in different forms with different intensity in different life cycles of the development of the cluster system. These differences establish the patterns, factors, conditions and constraints, as well as the impact of the level of synergy at each stage, and can be used to generate a multiplicative synergistic effect of the overall performance of the cluster system. For example, at the initial stage of development, a synergy can appear in the form of simplified components, e.g. in the form of reducing costs, saving time, etc.

Another stage in the formation of synergy is related to operational costs and investments, in particular, there can be distinguished: economies of scale, diversification of resources, growth directions, price elasticity of products, etc.

The next stage can be described in terms of 'qualitative' characteristics of synergetic advantages: mutual use of existing competencies and skills (the experience curve); strengthening of the brand and investment attractiveness; simplified access to strategic market segments, etc.

The accumulating stage can combine the results obtained at each stage and generate an overall synergistic effect. In this case, the cumulative systemic synergy combines the results of the development synergy effect at each stage and can be described by some function of the synergistic effect (for N stages): $F \supseteq \{(x^1{}_1, x^1{}_2, \ldots, x^1{}_n), (x^2{}_1, x^2{}_2, \ldots, x^2{}_k), \ldots, (x^N{}_1, x^N{}_2, \ldots x^N{}_m)\}$, where the parameters of the function F are the factors of synergy for a specific stage of the development cycle of the economic system: $(x^1{}_1, x^1{}_2, \ldots, x^1{}_n)$ – group of factors of synergy of the 1st stage of the life cycle of development of the economic system; $(x^2{}_1, x^2{}_2, \ldots, x^2{}_k)$ – group of factors in the 2nd stage of the life cycle; $(x^N{}_1, x^N{}_2, \ldots x^N{}_m)$ – group of factors in the nth stage of the life cycle.

In addition to taking into account the synergy of the different stages of the economic development life cycle, it is necessary to take into account the mutual influence of value chains (networks) operating within the framework of the economic system. At the same time, the interaction of participants in the value creation network requires the coordination and synchronisation of the management parameters in a unified development strategy, e.g.: management and operational technologies, allocated resources, interests of participants, procedure for harmonising tools and mechanisms of impact (activities, interests, resources, managerial influences).

The strategy for the sustainable development of the cluster network includes not only the active elements of the cluster system network representation that comprise the systemic development effect, but also mechanisms for generating development synergies in all possible directions of growth. The marginal value of the system effect will only appear if the components of development synergy meet the conditions and principles of development, ensuring consistency, balance and stability of strategic changes for managing the development of the cluster network.

6. Other Components of Development Synergy (Management Synergy)

Each basic synergetic component can in turn be structured into meaningful sub-components. Consider, for example, the component of the synergy of control functions ($\Delta^U E_S \subseteq \Delta E_S$), which can be decomposed into the following elements: planning $\Delta^U E_S^P$, forecasting $\Delta^U E_S^{\mathrm{Pr}}$, budgeting $\Delta^U E_S^B$, accounting $\Delta^U E_S^W$, analysis $\Delta^U E_S^A$, control $\Delta^U E_S^C$

$$\Delta^U E_S = \langle \Delta^U E_S^P + \Delta^U E_S^{\mathrm{Pr}} + \Delta^U E_S^B + \Delta^U E_S^W + \Delta^U E_S^A + \Delta^U E_S^C \rangle. \qquad (3)$$

Given that the synergistic effect changes in time, it is necessary to take into account the possible discounting of the synergy parameters using the discounting factors of the internal management processes (r_1) and the external environment (r_2); the synergy of management with discounting can be described by the following expression:

$$\Delta^U E_S^d = \Delta^U E_S \frac{(1+r_1)^n}{(1+r_2)^n} = \frac{(\Delta^U E_S^P + \Delta^U E_S^{Pr} + \Delta^U E_S^B + \Delta^U E_S^W + \Delta^U E_S^A + \Delta^U E_S^C) \cdot (1+r_1)^n}{(1+r_2)^n} \quad (4).$$

Similarly, it is possible to calculate the contribution to the systemic synergetic effect taking into account the recoil for a certain period of time for each synergetic component.

7. Conclusion

Thus, a systematisation of the factors and components of the synergetic effect in the development of the Arctic Cluster network, in which all circumpolar countries can participate, has been carried out. The conditions, types and types of the generated synergy components and possibilities for their analysis and evaluation have been demonstrated.

Acknowledgements

This paper is based on research carried out with the financial support of the grant of the Russian Science Foundation (Project No. 14-38-00009). Peter the Great St. Petersburg Polytechnic University.

References

[1] Efremova I *et al* 2017 Disparities in rural development of the Russian Arctic zone regions *Research for Rural Development* **2** 189
[2] Rudenko D Y, Pogodaeva T V and Didenko N I 2015 Poverty alleviation strategies in the Russian arctic zone regions *Mediterranean Journal of Social Sciences* **6 (1)** 32–39
[3] Romashkina G F, Didenko N I and Skripnuk D F 2017 Socioeconomic modernisation of Russia and its Arctic regions *Studies on Russian Economic Development* **28 (1)** 22–30
[4] Didenko N I *et al* 2017 Analysis of rural areas development of the region using the ADL-model *Research for Rural Development* **2** 142
[5] Yang Z, Hao P and Cai J 2015 *Cities* **42(B)** p171-185
[6] Li C, Leng Y and Sun J 2015 *Systems Engineering Procedia* **4** p 153-160
[7] Ivanova I, Strand Ø and Leydesdorff L 2017 *J Knowl Econ* Available from: https://doi.org/10.1007/s13132-017-0480-2 [Accessed 3 May 2018]
[8] Camagni R P, Diappi L 1991 *Regional Science* (Berlin, Springer)
[9] Mathews M D 2007 *Int Adv Econ Res* **13(2)** 254
[10] Brock W A, Xepapadeas A 2018 *Handbook of Enviromental Economics* Available from: https://doi.org/10.1016/bs.hesenv.2018.02.001 [Accessed 5 May 2018]
[11] Liu G Y et al 2012 *Ecological Informatics* **10** p 2-9
[12] Skripnuk D et al 2017 The formation of the regional tourist cluster in the Euro-Arctic Region *Int. Mult. Sc. Conf. on Social Sciences and Arts, SGEM 2017* **IV** 369-374
[13] Didenko N I et al 2017 Analysis of convergence-divergence in the development of economic processes in circumpolar countries *Int. Mult. Sc. Conf. on Social Sciences and Arts, SGEM 2017* 537-545

IOP Conf. Series: Earth and Environmental Science **180** (2018) 012012 doi:10.1088/1755-1315/180/1/012012

Economic and geopolitical aspects of developing the Northern Sea Route

N I Didenko, V I Cherenkov

[1] Peter the Great St. Petersburg Polytechnic University, St. Petersburg, Russia
[2] St. Petersburg State University, St. Petersburg, Russia

didenko.nikolay@mail.ru

Abstract. The present paper is devoted to define contemporary essence and logic of developing the Northern Sea Route (NSR) for the near future of Russia and beyond. A global nature of such upcoming processes as melting the Polar Ice Cap, depletion of readily available oil/gas reserves (causing a rise in prices on such commodities) and emerging confrontation between interests of circumpolar countries and non-circumpolar countries (on the background of a fast-growing economic and military power of latter and an ambiguity of legal status of using the Arctic space and its mineral resources) is defined. The features of Russian developing the NSR are historically and logically depicted and a decisive role of state political decisions and investments concerning the NSR is shown. The topological model of Russian Siberian-Arctic infrastructure grid where the NSR serves as its backbone is proposed. The said model and comparing data between domestic and transit turnover of cargo through the NSR have reinforced the hypothesis of the special importance of domestic functions of the NSR. A SWOT-matrix permitting to describe a current situation and future scenarios for the NSR is presented. The findings of the study allowed defining limitations and a vector of future research.

1. Introduction

Globalization has drastically impacted on the canvas of international trade corridors by increasing the density and dynamism of global merchandise flows not just between continents but also between West and East regions of the Eurasia continent. This *megacontinent* – located primarily in the Northern and Eastern Hemispheres and bordered by the Atlantic Ocean to the west, the Pacific Ocean to the east, the Arctic Ocean to the north, and by Africa, the Mediterranean Sea, and the Indian Ocean to the south – could be considered as an almost single entity – but in a geographical sense only. Factually, West and East had very different own histories in all dimensions – cultural, religious, language as well as social, political and industrial/military development ones. Such dissimilarities have created – in the course of development of East-West and West-East physical and virtual intensive communications – historical steady state logistics channels between European and Asian countries. However, due to the well-known – military-politically and ideologically confronted – geopolitical situation during seven decades of 20th century, the very shortest way between these countries (e.g., like the Silk Road) became impossible due to so-called “Iron Curtain” a phenomenon that existed until the end of the Cold War in 1991. The said phenomenon plus a general confrontation between “socialist and capitalist campuses” forced international trade logistics to by-pass overland routes between Northwest Europe and North-east countries of the Eurasia continent.

As it was calculated [1], over 80 per cent of world merchandise trade by volume and over 70 per cent by value was carried by sea and world seaborne trade volumes were estimated to grow at a compound annual growth rate of 3.2 per cent in 2017–2022. The same increasing trend existed in previous decades and expected to be continued in the near future. As they said, oceans did not separate people they connected them and did this by the cheapest way. Those who were skilled in the Art predicted that the Northern Sea Route, the NSR, (the said term was assigned as the English equivalent to *Severnyy morskoy put*, or *Sevmorput,* in Russian; but the same sailing corridor was often known outside Russia as the Northeast Passage, or the NEP, by analogy with the NSR) should be largely free of ice in the summers by 2050 if the Polar Ice Cap continued to melt at current rates [9]. So they began to discuss the economic attractiveness of the NSR as a shorter transportation corridor (table 1). However, the NSR was not considered as a real competitor for land transportation transit corridors, such as a "remake" of the historical Silk Road ("China the West" corridor that had been closed at first by Ottoman Empire boycotted trade with China and then by the "communism-capitalism" confrontation existed in the main part of 20th century); however, this "remake" received nowadays an additional option of Baikal-Amur Mainline for a future development and Russian Trans-Siberian railway, or the TSR (initially, Moscow-Vladivostok transportation arm), that could not become the main transportation land corridor for merchandise flows from Japan and Republic of Korea due to lasting till now splitting Korea into North and South ones and their political-military confrontation.

Table 1. Alternative Shipping Routes connecting Hamburg with East Ports, in nautical miles (*in comparison with the NSR*)

	From **Hamburg** to:			
Shipping routes via:	⇓	⇓	⇓	⇓
⇓	**Vancouver**	**Yokohama**	**Hong Kong**	**Singapore**
Northern Sea Route (NSR)	**6,635**	**6,920**	**8,370**	**9,730**
North West Passage*	N.A.	11,507	N.A.	N.A.
North West Passage / NSR	N.A.	*1,00*	N.A.	N.A.
Suez Canal	15,377	11,073	9,360	8,377
Suez Canal / NSR	*2,32*	*1,60*	*1,12*	***0,86***
Cape of Good Hope	18,846	14,542	13,109	11,846
Cape of Good Hope / NSR	*2,84*	*2,10*	*1,57*	*1,22*
Panama Canal	8,741	12,420	12,920	15,208
Panama Canal / NSR	*1,32*	*1,80*	*1,55*	*1,56*

Source: data from [17, p.1, table 1.1] (accessed May 18, 2018) were processed by authors.

The history of navigation on the NSR had its beginning somewhere in the mid of 17th century. However, the first chart showing how to cross the Arctic Ocean in this area was realized in the third quarter of 19th century. The very heavy Arctic ice was the main reason of losing ships and people of first Arctic expeditions. Initially, in Soviet time, the main driver for developing cargo shipping on the NSR was so-called "northern delivery" (*severny zavoz*) to supply the Far North regions of USSR with almost all merchandizes necessary to work and survive in these extremely important from economic and military viewpoints regions. Transit West-East-West commercial voyages for foreign freight were not considered (and even were banned) that time. Then, immediately after the WW2, the USSR had only a few not-so-powerful icebreakers, which were not able to secure the entire commercial voyages on the NSR. Despite the fact that in Soviet times (1964-1990) a numerous icebreaker fleet (4 units with power plant ≥ 45 000 BHP; 9 units with power plant < 45 000 BHP but ≥ 20 000 BHP; 12 units with power plant < 20 000 BHP but ≥ 10 000 BHP – *calculated by the authors using data from* [11]) has been built, the NSR was factually closed for foreign vessels due to military and political barriers. During and after so-called *pere-*

stroika, cargo shipping on the NSR collapsed in the absence of Russian state budget support. In spite of the fact that it was allowed foreign vessels to navigate on the NSR (1987), first real steps of revitalizing the NSR as an international transit transport corridor were revealed only from the year 2009. There were no questions that – under domination of state property in USSR – the Soviet state was a single political power and investor for developing the NSR having a well-understood strategic military and economic meaning should be better evaluated in our days. By the way, it was a reason of poor (backroom) statistics on NSR's development and maintaining that conserved its ambiguity till today; and a reason of a set of lacunas in the table 2 herein.

For contemporary Russia, the task of modernizing and converting the NSR in a reliable and economically efficient and almost year-round transportation corridor was defined as a vital task for building an integral social and economic infrastructure to engage a gigantic Siberian potential in the future development of Russia. The mega-grid of the said infrastructure could be formed by Siberian rivers crossed the country space between the South economically developed belt (formed along the TSR) and Russian Arctic regions (fig. 1). Modern transportation technologies (*e.g.*, "river-sea" boats, screen-plains, satellite communication means – including the Russia made GLONASS, a space-based satellite navigation system providing an alternative to GPS) and recent Russian government political decisions [2] [3] gave a realistic support for rather fast creating the said infrastructure.

Table 2. Some data on the cargo/passengers flows transited the NSR in years 2011-2016 in comparison with total volume of cargo transportation through the NSR.

Years	2011*	2012	2013	2014	2015	2016
Amount of vessels	26**	46	71	22	18	39
Rate to 2012 data		100	154	49	39	85
Quantity (metric tons)	N.A.	1 261 545	1 355 897	274 000	39 586	214 513
Rate to 2012 data	N.A.	100	108	N.A.	3	17
Gross Registered Tonnage/Tons	N.A.		1 397 051	463 928	N.A.	N.A.
Rate to 2013 data	N.A.	N.A.	100	33	N.A.	N.A.
Quantity of passengers	N.A.	N.A.	N.A.	N.A.	137	246
Average NSR passage time (days)***	N.A.	9,8	11,8	17,1	10,6	14,3
Total cargo transportation volume along the NSR	820 800	3 895 900	3 914 001	3 982 000	5 432 000	7 266 000
Share of the NSR's transit	N.A.	0,325	0,346	0,068	0,007	0,003

Notes:

* First year of a noticeable increase of foreign carriers to NSR transits while relevant Russian government programs began to revitalize the NSR since year 2009.

** Plus 14 vessels have been marked as "ballast"

*** These data were more susceptible to weather factors than to economic/geopolitical ones.

Sources: main set of data was collected from [8] (accessed May 18, 2018) and processed by authors; quantity (metric tons) for 2014 - The Rare Earth Magazine, JUN01, 2016; www.bellona.ru, APR09, 2014

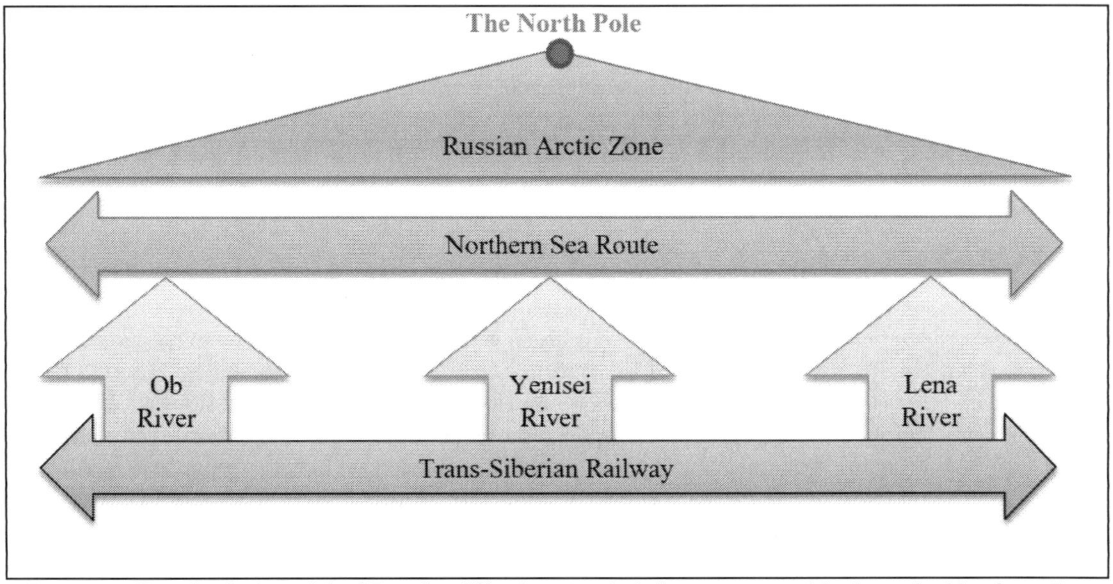

Figure 1. Topological model of Russian Siberian-Arctic infrastructure grid where the Northern Sea Route serves as its backbone

Source: designed by authors

2. Economic aspects of using the North Sea Route

The land-link between Asia and Europe – among the oldest trade routes in the world mentioned above – was for a rather long time not in use for any large-scale intercontinental trade; for containerized cargo especially. Hence, the total volume of international containerized cargo shipped using land transportation option between Asia (China mainly) and Europe was very limited. Some estimations assessed that the said rail transportation, in particular the TSR, could account for up to 3-4 % only of an average volume for 2014-2017 period, stemmed mainly from the North of China and the both Koreas. Besides, the distance between Northwest-European ports and the Far East was reduced approximately 40% by using the NSR as an alternative route compared to the traditional one through the relatively shallow Suez Canal (synonym: Maritime Silk Road, MSR). So, large-capacity vessels have been forced to circumambulate Africa in order to enter the Atlantic Ocean. Therefore, the NSR had for long attracted interest from observers, shipping companies and scholars because of its shorter distance and, hence, a time consumed (table 1) between Atlantic and the Pacific ports of origins/destination [5]. For comparison, according to some updated calculations [19] the great-circle distances taken from the GeoDist database from CEPII gave economies in terms of distance between ports of shipping (in Netherlands, United Kingdom, Belgium, and Germany) and ports of destination: Taiwan – 16-20%; China – 22-25%; South Korea – 30-33%; Japan – 36-39%).

Despite poor statistics on NSR traffics (due to several changes in defining official NSR's borders, careless monitoring and, perhaps, paranoid secrecy of Soviet period) certain statistics on sea ice concentration and ship traffic during a yearly cycle (table 2) permitted to formulate following considerations: (1) a share of (1) foreign participation in the NSR transit was geopolitically susceptible – e.g., after the Ukraine crisis began (2014) the more or less significant increase of the NSR transit was revealed in year 2016 only (however, there are additional nature and economic factors for decreasing the NSR transit at the same period – unfavorable changes in ice coverage and declining oil/gas prices, respectively); (2) international economic aspects of using the NSR became to prevail the geopolitical ones; indeed, for year 2017 the volume of NSR transit achieved 273 000 MT: (3) Russian government politics should be accentuated on domestic transportation tasks and success could be got in this strategic direction of developing the NSR should be the best PR for worldwide NSR's services offering.

They said [18] that currently the NSR was considered as a Russian internal sea route where international shipping was generally welcome but most likely without big hopes of being the only source to cover the huge investments necessary to modernize, maintain, and (potentially) expand the existing icebreaker fleet with new superpower units, infrastructure network along the NSR and corresponding satellite tracking/monitoring system. On the other hand, they should consider the NSR as a backbone of a gigantic infrastructure grid (fig. 1) supporting socio-economic development of offshore Arctic Ocean regions and Siberian ones where there are such obstacles for creating railroads and/or road networks as a harsh climate and an ambiguity of permafrost soils in the light of continuing global warming. There were two main intra-Russian traffics in the mainstream of the NSR: (1) *outbound one* – mineral resources, hydrocarbons, and timber for exports and domestic use and (2) *inbound one* – goods for life support, equipment, construction materials and other capital goods and, and it should be considered, defense goods. Consequently, while the NSR was of regional relevance and potential importance as a niche route [18] for a modest quantity of foreign transit goods on the background of much more significant and fast growing domestic and Russian export deliveries (*e.g.*, 2017 intra-Russia NSR traffic volume was 9,73 mln MT), the NSR's geopolitical and geo-economic relevance, especially in the sense of searching for global trade patterns, was estimated as rather limited. However, nowadays geopolitical factors played a decisive role and even prevailed over economic ones (e.g., the recent story with two *Mistral* helicopter carriers have been built but not handed over to Russian customer). Arctic region was many times called as a perhaps last unopened pantry of energy and other vital resources on the Earth and it was a good reason to consider Arctic as a focus of many geopolitical interests.

3. Geopolitical factors concerning the North Sea Route

Geopolitical position on the globe, optimistic estimations of energy resource potential and – in the frame of sustainable development concept – ecological significance of the Arctic region for the rest of the Earth have been drawing attention of the most part of nations on the eve of 21st century. The United States Geological Survey estimated (2008) that 22 % of the world's oil and natural gas could be located beneath the Arctic ice [20]. Even three-four decades before, they had greatly increased interest in the Arctic region not only from so-called circumpolar Nations or "Arctic Eight" (Canada, Finland, Denmark with Greenland and Faroe Islands, Iceland, Norway, Russia and USA due to Alaska), but other states located far from borders of the said region: other EU-countries and certain countries of the Asia-Pacific Region (first of all, China, India, Japan, Singapore, and South Korea). Along with using and building powerful icebreakers in non-Arctic countries (e.g., China, Germany, France), those interests in Arctic zone could be attested to by their observer status in the Arctic Council (China, France, Germany, India, Italy, Japan, Netherlands, Poland, Singapore, South Korea, Spain, Switzerland, United Kingdom). The resource and military importance of Arctic attracted nowadays such international organization as NATO. For Russia, began the development of the polar region eight centuries ago, the Arctic remains today an integral part of the Russian geopolitical space. Russian Arctic zone (RAZ) has been estimated as a set of the most important in economic and military sense regions of Russia. This zone made a regional contribution in the national income of Russia equal to 11% (with only 1% of country's population living therein) 11%/ And up to 26% of Russia's exports were stemmed from this zone [7]. Besides, the Arctic region was the only region where Russian and U.S. interests were vis-à-vis confronted in geographical sense. During the Cold War the naval (mainly, submarines) and air forces (strategic bombers) of both countries constantly patrolled over and under the Arctic ice fields, respectively. Nowadays, according certain sources, such situation could be in some ways considered as a recovered one. However, even after president Putin's Munich speech (2007) this viewpoint was not a common one. For example, it was noted [13] that the Norwegian government, fortified by the country's long history of peaceful co-existence with Russia considered the said situation as "'a return to a more normal level of activity for a major power with legitimate interests in the region" emphasizing that it could not be considered as Russian military developments directed against Norway.

In view of what has been said above, in the 2014 Military Doctrine of the Russian Federation, the Arctic was mentioned as a region where the Armed Forces must protect Russia's national interests even during peacetime. Contrary to the facts that Russia's policy was permanently declared as an appeal to cooperation some Western (mainly, American) politicians and journalists continued to accuse Russia striving to make its dominancy in Arctic. Indeed Russia, in the last decade, has increased a number of Arctic military forces, organized military drills, opened or reconstructed military bases, created powerful icebreakers, and established advanced radar stations to enhance its control of the said region. However, Russia had in Arctic zone (2010), in comparison with USSR (1980s) following levels of heavy armaments (calculated by authors using data from [3]): submarines – 0.17 (SSBN, Strategic Submarine Ballistic Nuclear – 0,18); aircraft carriers – 0,50; larger ships – 0,23; auxiliary vessels – 0,16; aircrafts – 0,25. Such a sharp decline in these levels may indicate that many of Russia's military-related activities are pragmatic and cooperative in nature – minimally necessary defensive measures. As the zone of the NSR becomes more suitable for economic development, Russia was improving its radar surveillance and communication therein. The same could be said about developing long-range radar systems and drone bases that were destined to enhance control and surveillance over Russia's oil reserves and maritime shipping. For contemporary Russia, inherited after turbulent 1990s almost open Northern state border, the Arctic remained an integral part of the Russian geopolitical space, both in global economic and military sense.

4. Some notes concerning developing the North Sea Route

A remarkable foresight for a future exploiting the NSR could be found, for example, in outputs of the joint research project "Feasibility and Reliability of Shipping on the Northern Sea Route and Modeling of an Arctic Marine Transportation & Logistics System" [3]. A remarkable feature of the said project was presented not only due to high professional levels of corresponding participants but due to their geographical positions regarding to the topology of the NSR – two edges (Norway and South Korea) and the main course (Russia).

Russian resource extracting industries were permanently faced – along the NSR and in Russia's Arctic zone (RAZ) as a whole – with heavy operational sea/land conditions including: (1) a short drilling season, (2) very long inbound and outbound supply chains, (3) extreme cold temperatures most of the year, (4) storming, (5) thick icing (sometimes, up to 4 meters), (6) full darkness during so-called "Polar night" (lasing about one hundred days), (6) fast changing sea-ice conditions, (7) heavy fogs, (8) offshore operations in deep waters, and (9) increased coastal erosion and permafrost thawing in the summer impacting land-based infrastructure by destabilizing foundations.). Almost all of these factors could be mitigated or even minimized under the condition of well-organized and technologically supported logistics of the NSR. Nevertheless, as it was announced in the Annual Presidential Address to the Federal Assembly (2018) Vladimir Putin said [15]: "The Northern Sea Route will be the key to developing the Russian Arctic and Far East." Taking into account several precedents (2014 – Sochi Olympics; 2018 – Crimean Bridge) they should seriously and positively estimate perspectives of future developing the TSR.

For contemporary Russia, the task of modernizing and converting the NSR in a two-way reliable and economically efficient transportation West-East-West corridor was defined as a vital task for building an integral social and economic infrastructure to engage a gigantic Siberian and Far East regions potential in the future development of Russia. The mega-grid of the said infrastructure (fig. 1) could be formed by Siberian rivers (Ob, Yenisei and Lena) crossed the country space – very rich with mineral and forest resources but having an undeveloped infrastructure; partly, due to between the South economically developed belt (formed along and near the TSR) and Russian Arctic regions. Modern transportation technologies (*e.g.,* "river-sea" boats, screen-plains, satellite communication means – including the Russia made GLONASS, a space-based satellite navigation system providing an alternative to GPS) and recent Russian government political decisions [2] [3] gave a realistic support for rather fast creating the said infrastructure.

As well-known, the classical definition of sustainable development was formulated as such development of that meets the needs of the present without compromising the ability of future generations to meet their own needs. This general formula has found its realization in directions of the socio-economic development of the Russian Arctic region in such ways as [16]: the social infrastructure, the institutional sector, the human development, the industrial sector and the ecology. As it was stated in the "Basics of the State Policy of the Russian Federation in the Arctic for the Period till 2020 and for a Further Perspective" (hereinafter referenced as the "Basics"), low sustainability of ecological systems determining the biological balance and climate of the Earth, and their dependence even on insignificant anthropogenous impact was one of specificities of the Arctic zone of the Russian Federation to be taking into account in formatting the Russia's Arctic policy. Corresponding items could be found in programs of corporative social responsibility elaborated by and to be followed by so-called natural monopolies (e.g., *Gazprom*) that could be estimated along with the state as powerful drivers of developing the NSR and RAZ as a whole.

5. Considerations and limitations

After reviewing a substantial set (about 60) of recently published academic articles as well as reports and statistics concerning the NSR and issued by corresponding national and international organizations it became possible to make some considerations concerning the near future of the NSR in the light of economic, natural, technological and geopolitical factors. A summarized authors' vision of issues discovered by the present article was displayed herein in the SWOT-matrix format (table 3). In addition some comments in brief were presented in the following text. First of all, almost all of factors mentioned above, one way or another, could be correlated with transportation risks of different nature.

Unfortunate combination of Arctic climate conditions, accidental events in aquatorium of the NSR, and processes in the relevant logistics management systems were recognized as the leading contributors to risks of ship accidents. The contemporary risk analysis [6] – done to make pilots and captains more experienced in conducting ships through the NSR – revealed that the East Siberian Sea had the highest probabilities regarding collision, foundering and grounding of the ship. Other regions such as Chukchi, Laptev, Kara and the Barents Sea had almost similar probabilities regarding grounding. The same time, foundering probabilities were very low in the said areas. Such knowledge helped to improve the risk management system on ships passing through the NSR.

In addition to improving these risk management systems on micro-level (vessels) it was defined as necessary to arrange measures for implementation of integrated approach to public and territorial safety in the RAZ for regions of future intensive and sustainable development in the context of expected climate changes (global warming) and other disasters include the following [7]: (1) establishing a network of integrated centers with the National Crisis Management Center as an informational hub connecting relevant regional and federal services and ministries for corresponding emergency monitoring, forecasting, and warning; (2) improving and practicing the ways and procedures of sharing the capabilities Russian System of emergency prevention and response (RSCHS); (3) assigning relevant Atomflot-operated vessels to operations of RSCHS emergency response forces; (4) arranging additional storage bases well-equipped for effective rescue operations; (5) developing innovative and improving existing technologies for rescue and other urgent operations at low temperatures; (6) using/improving applications of Arctic-conditions-adapted rescue aircrafts and satellite-based systems for rescue operations and monitoring (KOSPAS-SARSAT, *Gonets* and – in the future – *Arktika*).

In conclusion, while defining a future of the NSR, some comments to concentrated SWOT-analysis of NSR logistics services offers s (table 3) should be made. First of all, new basics for future development of the NSR as a part of the RAZ should be elaborated. Then, as it seemed to the authors, while elaborating these new state directives it should be better to consider the NSR not only as a vital part of the RAZ but as a part of wider "infrastructure grid" displayed above (fig. 1) as a topological model of Russian Siberian-Arctic infrastructure grid where the Northern Sea Route serves as its backbone. This statement could be backed by the principle of preserving integrity of product logistics flaws flawing in and out of the NSR. Such approach – from the very beginning, on the stage of conceptual designing new basics – could help to avoid/mitigate risks of additional logistics delays/disruptions could take place due to mismatching in "sea-river" interfaces.

IOP Conf. Series: Earth and Environmental Science **180** (2018) 012012 doi:10.1088/1755-1315/180/1/012012

Table 3. SWOT-matrix for the NSR logistics services offers

STRENGTHS	WEAKNESSES
→ Reduced sailing time defining reduced production costs (fuel consumption, depreciation of vessel, salaries, etc.) → The most numerous and the most powerful in the world Russian fleet of icebreakers → Specialized RF government programs on developing production/military/social infrastructure adjoining the NSR	→ Partly shallow waterway → Sophisticated administrative procedures → Underdeveloped circumlittoral logistics infrastructure along the NSR → Increased insurance premia → The predominantly one-way traffic with ballast (without a back-haul) → Icebreaker made channels are narrower than width of *Panamax/Snesmax* class tankers
~ *Supports from resource extracting companies (so-called natural monopolies)* ~ *PR as a result of international research and conferences* ~ *Less CO2 global emission* ~ *Lack of sea piracy along the NSR*	~ *Severe climate and ice conditions* ~ *Increased nature and operation ("Arctic influenced human factor") risks* ~ *Some uncertainty in the legal regime of using the NSR* ~ *Certain vulnerability to geopolitical risks*
OPPORTUNITIES	**THREATS**
→ New growth of oil/gas prices → Further global warming as a factor of increasing a time of Arctic navigability → Innovative solutions for Arctic sea logistics (*e.g.*, satellite tracking ships/cargo, holistic digitalizing) → Substantial growth of Russian oil and gas transportation (*e.g.*, due to fast development of the Yamal LNG project)	→ Substantial improvements in competitive advantages of the NWP → Expected increasing number of sea incidents due to more intensive traffic of oil and gas tankers → Emerging competitors offering Northern transportation routes out of Russian exclusive economic zone (NWP or Transpolar Way)
~ *Strategic alliancing with companies from Arctic as well as non-Arctic countries* ~ *Weakening a competitive advantage of the MSR due to an aggravation of geopolitical situation inspired by international territorial disputes of China and its confrontation with U.S. (Malacca and Suez bottlenecks)* ~ *Entering in force the Polar Code* ~ *More clear Russian legislation and administrative procedures concerning the NSR users*	~ *Further world economic recession* ~ *New economic and political counteractions against Russia's expansion in Arctic zone* ~ *Revitalizing international waters disputes* ~ *Exacerbating geopolitical "soft power" actions against "Russian Arctic expansion"* ~ *Exacerbating actions and aggressive media politics of Greenpeace or similar NGO* ~ *Hidden counteractions from the side of non-Arctic shipbuilding industry stakeholders*

Legend:

→ Direct impact

~ *Indirect impact*

Source: compiled by the authors on the basis of some sources listed in References herein.

In conclusion, while defining a future of the NSR, some comments to concentrated SWOT-analysis of NSR logistics services offers (table 3) should be made. First of all, new "Basics" for future development of the NSR as a part of the RAZ should be elaborated. Then, as it seemed to the authors, while elaborating these new state directives it should be better to consider the NSR not only as a vital part of the RAZ but as a part of wider "infrastructure grid" displayed above (fig. 1) as a topological model of Russian Siberian-Arctic infrastructure grid where the Northern Sea Route serves as its backbone. This statement could be backed by the principle of preserving integrity of product logistics flaws flawing in and out of the NSR. Such approach – from the very beginning, on the stage of conceptual designing new "Basics" – could help to avoid/mitigate risks of additional logistics delays/disruptions could take place due to mismatching in "sea-river" interfaces.

Naturally, being understood a role of the Arctic in the planetary ecology the authors came to conclusion that the concept of sustainable development had to thread many articles of expected "Basics". The mode of NSR functioning should be of such style that permitted not only to ensure outbound logistics flows (crude oil, LNG, timber, ore, etc.) and inbound logistics flows (capital goods, vital resources, etc.) but realize tasks of developing sustainability logistics. The latter could comprise of implementing innovative eco-friendly technologies; such as, nuclear icebreakers, movable nuclear power blocs, waste recycling plants, etc. should include implementing well-branched network of ground/space weather stations, multichannel communication lines, and IT-technologies permitting to manage NSR's traffic in as much effective and efficient as possible way. The "Basics" should twist together business, social and ecological decisions (so-called 3P-paradigm – Profit, People and Planet) in such a way that was in good conformity with the main concept of sustainable development [16]. For this purpose, not only investments and technologies but NSR focused education for sustainability should be provided.

It would be useful to stipulate in the "Basics" a primacy of domestic and Russian export tasks to be resolve with help of the NSR. Positive outputs of the said tasks resolved should become the best evidence of well-arranged, effective and efficient infrastructure of the NSR. In this case, the NSR's transit should be rather easily converted from buzzing topic of today (compare: one day Suez traffic was approximately equal to an annual one of the NSR) into a real competitive transportation option. The said option should be supplied with well-developed coastal logistics infrastructure (navigation, pilotage and rescue services), space tracking/monitoring system, and powerful icebreakers. As a result, geopolitical confrontations and emerging Arctic territorial disputes should be attenuated by economic and financial considerations stemmed from expected successful functioning of the NSR. Besides, an investment attractively should be augmented due to same reasons.

References

[1] UNCTAD 2017 *Review of Maritime Transport 2017* United Nations publication sales No E.17.II.D.10, New York and Geneva.

[2] Russia's Arctic Strategy: Aimed at Conflict or Cooperation? (Part I) 2018 Available from: https://www.thearcticinstitute.org/russias-arctic-strategy-aimed-conflict-cooperation-part-one/ [Accessed 11th May 2018]

[3] Russia's Arctic Strategy: Military and Security (Part II) 2018 Available from: https://www.thearcticinstitute.org/russias-arctic-military-and-security-part-two/ [Accessed 11th May 2018]

[4] Research: CHNL's Arctic 2030 Project 2015 On-line URL: http://www.chnl.no/?page=4&show=90&news=68&title=Research%3A+CHNL%92s+Ar Available from [Accessed 11th May 2018]

[5] Lasserre F 2014 Case studies of shipping along Arctic routes Analysis and profitability perspectives for the container sector *Transportation Research Part A* **66** 144-161

[6] Baksh A-A Abbassi R Garaniya V and Khan F 2018 Marine transportation risk assessment using Bayesian Network: Application to Arctic waters *Ocean Engineering* 159 422-436

[7] Akimov V A Sokolov Yu I 2010 Risks of Emergencies in Russia's Arctic Zone *Issues of Risk Analysis* **7** **(4)** 4-24

[8] NSR Transit Statistics *CHNL Information Office Northern Sea Route O* Available from:http://www.arctic-lio.com/nsr_transits [Accessed 11th May 2018]

[9] Aksenov Y *et al* 2017 On the future navigability of Arctic sea routes: High-resolution projections of the Arctic Ocean and sea ice *Marine Policy* **75** 300–317
[10] International Code for Ships Operating in Polar Waters (Polar Code) 2017 Available from: http://www.imo.org/en/MediaCentre/HotTopics/polar/Documents/POLAR%20CODE%20TEXT%20AS%20ADOPTED.pdf [Accessed 11th May 2018]
[11] Melia N Keith Haines K and Hawkins E 2017 Implications from Opening Arctic Sea Routes Foresight *Future of the Sea Evidence Review - Foresight, Government Office for Science*
[12] Marchenko N A *et al* 2018 Arctic Shipping and Risks: Emergency Categories and Response Capacities *TransNav the International Journal on Marine Navigation and Safety of Sea Transportation* **12 (1)** 107-114
[13] Blunden M 2009 The New Problem of Arctic Stability *Survival* **51 (5)** October–November 121–142
[14] Dean J W 2018 NEO The Huge Implications of Russia's Northern Sea Route *Veterans Today* January 3 Available from: https://www.veteranstoday.com/2018/01/03/neo-the-huge-implications-of-russias-northern-sea-route/ [Accessed 11th May 2018]
[15] Presidential Address to the Federal Assembly 2018 Available from: http://en.kremlin.ru/events/president/news/page/106 [Accessed 11th May 2018]
[16] Didenko N, Kunze K and Skripnuk D 2015 Analysis of the development in the Russian Arctic zone *Sustainable Development and Planning* Available from: https://www.witpress.com/elibrary/wit-transactions-on-ecology-and-the-environment/193/33837 [Accessed 11th May 2018]
[17] Ragner C L 2000 Northern Sea Route Cargo Flows and Infrastructure Present State and Future Potential *Fridtjof Nansen Institute Report* 13/2000 Available from: https://www.fni.no/publications/northern-sea-route-cargo-flows-and-infrastructure-present-state-and-future-potential-article717-290.html [Accessed 11th May 2018]
[18] Raspotnik A and Stephen K 2013 Why the Northern Sea Route is Still of Limited Geo-Economic Importance The Arctic Institute Center for Circumpolar Security Studies Report SEP08 2013 Available from: https://www.thearcticinstitute.org/northern-sea-route-limited-geo-economic-importance/ [Accessed 11th May 2018]
[19] Bekkers E Francois J F and Rojas-Romagosa H 2018 Melting Ice Caps and the Economic Impact of Opening the Northern Sea Route *The Economic Journal* **128 (610)** 1095-1127
[20] CircumArctic Resource Appraisal: Estimates of Undiscovered Oil and Gas North of the Arctic Circle *USGS Fact Sheet,* 2008-3049 2008 Available from: http://pubs.usgs.gov/fs/2008/3049/fs20083049.pdf [Accessed 11th May 2018]

Environmental Shocks: Modelling the Dynamics

N I Didenko[1] **and S V Kulik**[12]

[1]Peter the Great St. Petersburg Polytechnic University, St. Petersburg, Russia
[2]St. Petersburg State University, St. Petersburg, Russia

didenko.nikolay@gmail.com

Abstract. This paper reports on processes occurring within the human environment: shocks occurring in the environment; the increase in greenhouse gases from industry; the increase in greenhouse gases from agriculture; the change in global temperature; the reduction of freshwater resources; the change in area of forested land. The article also examines the impact of three groups of factors on the dynamics of environmental shocks. These factors are the history of the process, the impact of various dynamic factors of human life and the degree of influence of the processes occurring in the environment. The autoregressive distributed lags model (ADL model) is chosen as the theoretical framework for the analysis of the correlation between the environmental shocks and the influencing factors, and empirical testing of the model is performed. The exogenous variables characterising the factors of human life were analysed using the following indices: global GDP per capita; economically active global population; global merchandise exports (FOB); rail transport freight turnover (mln tons km^{-1}); the mining of coal and lignite, extraction of peat, (mln USD). The forecast of the development of environmental shocks in 2017-2018 was performed. The analysis of the model showed that the dynamics of the environmental shocks had a constant trend.

1. Introduction

The environment comprises the totality of natural conditions, the objects of animate and inanimate nature and the components of the human (as well as animal and plant) environment. In this way, the environment's form and nature can be defined [1]. The environment surrounding mankind is continuously changing. The climate is changing, pollution of the environment is taking place and both forested area and freshwater resources are declining.

The consequences of the occurrence of one of these processes, referred to as "environmental shocks", are investigated in this article. The shocks consist of those disasters that sometimes occur in the environment. The shocks occurring in the environment are estimated in terms of their economic damage. The economic damage caused by environmental shocks is growing rapidly every year. In the 1960s it amounted to about 1 bln USD, in the 1970s – up to 4.7 bln USD, in the 1980s – 16.6 bln USD, in the 1990s – 76 bln USD [2].

Economic losses from natural shocks add to the damage caused by the following factors of human activity on the planet: water pollution; soil pollution and degradation; acid rains; ozone depletion; waste; loss of biodiversity. Concerning waste production activities, significant economic damage to society is caused by the methods of waste disposal – landfilling or incineration. These methods are economically unprofitable and cause great economic losses. A pressing problem of humanity today is the loss of biodiversity. Annually about 15,000 species become extinct. Extrapolating from this information, in 50 years only half of the current biological diversity of our planet will be conserved [3]. It is impossible to use economic damage to quantify the process of biodiversity loss

The relevance of the problem under investigation is presented in the analysis of the environmental shocks. The aim of the study is to analyse the influence of various factors on the dynamics of the change in shocks, model the influence of human activities on environmental shocks and develop the exogenous variables effect model that reflect human activity on the endogenous variable that are the environmental shocks.

2. Literature Review

The study of the ecological problems is widely researched.

Classification of ecosystems is arranged according to the organisational form described in this paper [4]. It is of note that ecosystems can be defined in various ways, including or excluding humans. There are two different approaches to the classification of ecosystems: a deductive approach, based on the theoretical framework, and an inductive approach, based on the empirical data. Thus, human activities affect the ecosystems and at the same time they meet social needs.

The research, devoted to problems of origin and the evolution of ecosystems, is of interest. The study [5] introduces the concept and explores the environment from the position of the creation and assessment of ecological communities and ecosystem functions. The probabilistic scenarios of natural disasters are discussed. The paper [6] studies the impact of large-scale disasters on economic development. The night-time light intensity satellite data is chosen as a factor of economic development. It is shown that this indicator correlates strongly with income per capita. It is noted that natural disasters significantly reduce the amount of light visible from space in the short term. Climatic and hydrological disasters cause significant reduction in brightness in developing and emerging market countries. The geophysical and meteorological disasters in industrialised countries further weaken the intensity of illumination.

The factors affecting the size of the disaster are defined as the geographical location, the degree of financial development and the quality of the political institutions present. The analysis of the environmental trends alongside the development of human society is developed in this research [7-9]. Researchers formulate the endogenous growth model showing how environmental factors, corruption, public investment and military expenditure are linked [7]. This research [8] proposes a set of optimal taxes on emissions that can be used for the internalisation of external effects from environmental pollution. The study provides evidence for the proposition that production in low-income countries will be less adverse in terms of taxes on carbon emissions compared to the countries with a high level of income, whether replaced or not. The trade-off between economics and the environment is demonstrated and a management model for the sustainable development of the marine environment proposed. The adaptive management strategy in the conditions of the increasing rate of environmental change is presented [9].

This paper [10] proposes a methodological approach in order to assess the impact of economic activities on natural disasters. The article presents an input-output model of direct and indirect socio-economic impacts on natural disasters. The following indicators are considered socio-economic: population density, the level of employment and the turnover and the size of the share capital of companies.

The modelling of environmental and socio-economic effects is discussed in works [11-13]. The models describe cumulative effects that are best managed on a regional scale. However, most models of the environment presented in this set of scientific literature, provide local models for the environmental assessment at the level of different projects. Researchers justify the necessity of determining the overall risk to humans and the environment [14]. The Integrated Risk Assessment (IRA) as a combination of Environmental Exposure assessment (EEA) and Human Exposure assessment (HEA) is provided [15]. A model for uncertainty or risk assessment can also be constructed for situations when there is a danger of something having an impact on the environment. Processes occurring in the environment are primarily analysed from the position of the Earth's natural resources, which can be connected to human activities. The methodology for water resource assessment is presented as part of the planning, development and management of water resources at a national level. The factors of climate change, the global population growth and the rapidly changing patterns of land use change deep river basins. There exists an urgent need for further studies that would assess the risks of exploitation of water resources.

Demand for minerals is increasing world-wide as the population increases and the consumption demands of individual people increase. The mining of earth's natural resources is, therefore, accelerating, and it has accompanying environmental consequences [17]. Currently, companies need large amounts of various minerals, and this demand is constantly growing [18]. The issues concerning conservation of natural resources are under study. Ecohydrological modelling of water resources and land is used to assess the effectiveness of the use of water, land and forest resources for agricultural purposes [19] and forecast estimates are given.

The risks to the environment are also evaluated from the position of the life cycle of plants. This paper presents the model used to analyse the interaction between the life cycle of plants and the environment [20]. One way to understand the extent of the human impact on the Earth is the ecological footprint (EF). It is said that on a global scale EF exceeded the biocapacity of the Earth, i.e. the Earth's ability to produce resources and ecological services, in 1970. In 2012 the EF indicator increased to 1.6 times that of earth's capacity. However, the biocapacity of the Earth itself slightly increased during this period because of the increase in agricultural productivity [21].

In the current conditions of the development of the global environment it is necessary to develop a system for managing ecosystem services in the context of a sustainable global development policy [22-24], and robust monitoring mechanisms are required.

3. The influence of different factors on the development of environmental shocks

This article formulates the hypothesis that the development of environmental shocks is influenced by three groups of factors. The first group of factors is the history of the process, the second is the influence of various dynamic factors of human life and the third is the degree that the processes influence the environment.

Prehistory precedes everything

Dynamic factors of human life are the areas in which humans carry out various kinds of activities: social, productive, economic, demographic, environmental management. Human activity in these areas is estimated by the set of indicators that will be considered "exogenous indicators". Let us name some of them: global gross domestic product per capita; global employed population; global economically active population; production energy, utilities and recycling: (days sales outstanding); primary materials forestry – production (turnover); global road freight traffic; mining of coal and lignite; extraction of peat, and others.

The factors of human activities are developing to meet the needs of the inhabitants of the planet; this development generates adverse environmental consequences and initiates the development of environmental shocks.

The processes occurring in the environment affect environmental shocks. In other words, the environmental shocks depend on: the changes in the volume of CO2 in the environment; greenhouse gas emissions from industries into the environment; emissions of greenhouse gases from agriculture into the environment; changes in global temperature on the planet; freshwater resource reduction; change in forested area

4. Methodology for the mathematical modelling of correlations between environmental shocks and their influencing factors

Under the term mathematical modelling, we understand (a) the development of the mathematical model, i.e. a view in the form of a mathematical equation or a system of equations or arithmetic ratios of the most important properties of the real process, (b) the description of the algorithm for finding the mathematical model parameters, (c) the study of a real process in the model, i.e. finding the model parameters under the given change of the characteristics of the process.

As the theoretical framework of analysis of correlation between the environmental shocks and their influencing factors, the model of autoregressive distributed lags (ADL model) is chosen, in which the current values of the series depend on the past values of the series as well as on the current and past values of other time series.

The ADL model describes the hypothesis of the effects of exogenous variables upon the endogenous variable effectively. This model considers the current and prior periods and can be represented both in additive and multiplicative forms.

The model is generalised in the case of multiple exogenous variables. In the general case, it can be assumed that all exogenous variables are included in the model with the same number of lags.

The additive form of the ADL model is:

$$y_t^1 = a_0 + \sum_{i=1}^{n} a_i y_{t-i}^1 + \sum_{j=1}^{5} b_j X_t^j + \sum_{s=2}^{7} c_s y_t^s + E_t \qquad (1)$$

where $n+5+7$ is the number of exogenous variables;
n is the number of lags of the endogenous variable (i=1, 2,..n);
E_t - are residues that make up the process of white noise.
j=1, 2 ,..5 is the number of exogenous variables – dynamic factors of human activities;
s=2, …, 7 is the number of exogenous variables of the processes occurring in the environment;

The multiplicative form of the ADL model is:

$$y_t^1 = A_0 \prod_{i=1}^{n} y_{t-1}^{1^{\alpha_i}} \bullet \prod_{j=1}^{5} X_t^{j^{\beta_j}} \bullet \prod_{s=2}^{7} y_t^{s^{\delta_s}} + E_t \quad (2)$$

Taking the logarithm of the left and right part of the equation on the basis of e we obtain a linear equation of the form:

$$\ln y_t^1 = \ln A_0 + \sum_{i=1}^{n} \alpha_i \ln y_{t-i}^1 + \sum_{j=1}^{5} \beta_j \ln X_t^j + \sum_{s=2}^{7} \delta_s \ln y_t^s + E_t \qquad (3)$$

Thus, a multiplicative function can also be additive. The endogenous variable in the model (y_t^1) represents the environmental shocks. The first group of exogenous variables is the characteristic of the prehistory of the process, i.e. shocks that occurred in the environment in previous time periods (y_{t-j}^1). The second group of exogenous variables is the factors of human life activity, as measured by indicators: total global GDP per capita (USD per person) (X_t^1); the number of economically active people in the world (X_t^2), global merchandise exports (FOB) (X_t^3); the rail transport freight turnover, (mln tons km^{-1}) (X_t^4); the mining of coal and lignite, extraction of peat, USD mln (X_i^5). The third group of exogenous variables is the impact of the processes occurring in the external environment (on the shocks): a change in the content of CO2 in the environment (y_t^2); the growth of greenhouse gases from industry (y_t^3); the growth of greenhouse gases from agriculture in the environment (y_t^4); global temperature change (y_t^5);freshwater resource reduction (y_t^6); the change in forested area worldwide (y_t^7).

The algorithm for finding the mathematical model parameters involves a number of stages: collecting and processing of endogenous and exogenous primary data in the form of time series; validation of time series variables for stationarity using Dickey-Fuller test; checking the exogenous variables for multicollinearity; selection of the lags of the endogenous variable, which have a strong correlation with the variable value in the last period and testing the significance of coefficients of autocorrelation using the Ljung-Box Q-test; the choice of exogenous variables that have strong correlation with the value of the endogenous variable in the last period and testing the significance of correlation coefficients; the design of the structural form of the multiplicative form of the ADL models; working out the structural form of the ADL model in an additive form; determination of model coefficients using the least squares method; verifying the significance of the regression equation and coefficients of the regression equation.

Finding the model parameters under the given changed characteristics of the process consisted of predicting the endogenous variable for a specific set of time periods. Specifically, a forecast for the development of environmental shocks in 2017-2018 was made.

5. Empirical testing of the model

The time series data covering the time period from 1997 to 2016 is used as a source of information for developing the econometric function for statistical studies. Source: Euromonitor International, http://www.euromonitor.com/; World Bank Open Data http://data.worldbank.org/.

The indicator "change in global temperature, annual growth in %" represents the arithmetic mean value of all countries in each period. All the other parameters represent the sum value of all countries for each year. In the analysis the average values were used.

Testing the time series variables for stationarity is the solution, which is calculated independently for each variable, using the autoregressive equation in the form $y_t = ay_{t-1} + \varepsilon_t$, where y_t is a time series, ε_t is an error. If $|a| < 1$, the series is stationary. The verification of the coefficient reliability was carried out using t - statistics for the significance level $\alpha=0.05$.

For the endogenous indicator "environmental shocks" $\alpha=0.050 < 1$, $t_{calculate}=2.203 > t_{table}=1.729$.

For the exogenous indicator of "global GDP per capita" $\alpha=0.961<1$, $t_{calculate}=18.709 > t_{table}=1.729$.

For the exogenous indicator "rail transport freight turnover" $\alpha=0{,}960<1$, $t_{calculate}=54.096 > t_{table}=1.729$.

For the exogenous indicator "economically active global population" $\alpha=0.956 <1$, $t_{calculate}=65.611 > t_{table}=1.729$.

For the exogenous indicator "global merchandise exports" $\alpha=0.412 <1$, $t_{calculate}=1.895 > t_{table}=1.729$.

For the exogenous indicator of "mining of coal and lignite, extraction of peat" $\alpha=0.409 <1$, $t_{calculate}=1{,}849 > t_{table}=1.729$.

For the exogenous indicator of "CO2 Emissions per production unit" $\alpha=0.935 <1$, $t_{calculate}=47.02 > t_{table}=1.729$.

For the exogenous indicator "Greenhouse gas emissions from industry" $\alpha=0.355<1$, $t_{calculate}=1.860 > t_{table}=1.729$.

For the exogenous indicator "Greenhouse gas emissions from agriculture" $\alpha=0.818<1$, $t_{calculate}=10.709 > t_{table}=1.729$.

For the exogenous measure of "Global temperature change" $=0.209<1$, $t_{calculate}= 2.462 > t_{table}=1.729$.

For the exogenous indicator "Freshwater resource reduction" α=0.664<1, $t_{calculate}$ =6.429 > t_{table} =1.729.

For the exogenous indicator "Change in forested area" α=0.767<1, $t_{calculate}$ =8.509 > t_{table} =1.729.

The testing of the exogenous variables for multicollinearity was conducted by calculating the pairwise correlation coefficient between pairs of exogenous variables and removing one of the variables from the pair if the correlation coefficient was greater than |0.7|. The significance of correlation coefficients was assessed by Student's t-criterion for the significance level α=0.05.

As a result of this verification, the variables "CO2 emissions per production unit", "greenhouse gas emissions from agriculture" and "freshwater resource reduction" were removed from further analysis.

Choosing the endogenous variable lags, which have a strong correlation with the variable value in the current period, was carried out by computing the autocorrelation coefficients and testing the significance of coefficients of autocorrelation using the Ljung-Box Q-test. The autocorrelation analysis showed that the endogenous variable of the current period had an autocorrelation coefficient value of more than 0.7 with an endogenous variable of the previous period only. The autocorrelation coefficient is significant according to the value of Q criterion of Ljung-Box

$$Q = n(n+2)\sum_{k=1}^{m}\frac{\rho_k^2}{n-k} \qquad (4)$$

where n is the number of observations ρ_k is the autocorrelation of the kth order, m is the number of tested lags $Q = 20\cdot 22\left(\frac{0.734^2}{20-1}\right) = 12.496 > \chi^2_{1-\alpha,n} = 10.851$

Choice of exogenous variables that have a strong correlation with the value of the endogenous variable in the previous period and testing the significance of correlation coefficients.

The exogenous variables $y_t^3, y_t^5, y_t^7, X_t^1, X_t^2, X_t^3, X_t^4, X_t^5$ in the previous period have a strong impact on the endogenous variable in the previous period. The significance of the correlation coefficients was assessed by Student's t-criterion at the significance level of α=0.05 and the number of degrees of freedom n-2=16, t table=1.746. In all cases $t_{calculate} > t_{table}$. The correlation coefficients of other exogenous variables with the endogenous variable in the previous period are less than 0.7, therefore, they will not be included in the further analysis.

The structural form of the ADL model in multiplicative form is written as

$$y_t^1 = A_0 \cdot {y_{t-1}^{1}}^{\alpha_1} \cdot {X_t^{1}}^{\beta_1} \cdot {X_t^{2}}^{\beta_2} \cdot {X_t^{3}}^{\beta_3} \cdot {X_t^{4}}^{\beta_4} \cdot {X_t^{5}}^{\beta_5} \cdot {y_t^{3}}^{\delta_3} \cdot {y_t^{5}}^{\delta_5} \cdot {y_t^{7}}^{\delta_7} \quad (5)$$

Taking the logarithm of both parts of the equation allows linearisation of the function's variables and exponential model. This allows us to write the structural form of the ADL model in an additive form

$$A_t^1 = \alpha_0 + \alpha_1 A_{t-1}^1 + \beta_1 B_t^1 + \beta_2 B_t^2 + \beta_3 B_t^3 + \beta_4 B_t^4 + \beta_5 B_t^5 + \delta_3 C_t^3 + \delta_5 C_t^5 + \delta_7 C_t^7 \ (6)$$

By means of the least-squares determined coefficients model we check the significance and coefficients of the regression equation using Fisher's F-criterion and Student's t-criterion.

When α = 0.05, F_{table} = 30.144, $F_{calculate}$ = 52.844, the coefficient of determination equals 0.932; thus, the regression equation is significant.

The regression equation in an additive form is written as

$$A_t^1 = -30{,}935 + 0.24 A_{t-1}^1 + 1{,}997 B_t^1 - 2{,}867 B_t^2 + 1{,}163 B_t^3 + 0{,}427 B_t^4 - 0{,}321 B_t^5 - 0{,}380 C_t^3 - 0{,}185 C_t^5 + 0{,}085 C_t^7$$ [TB1](7)

The regression equation in a multiplicative form is

$$y_t^1 = e^{-30{,}935} \cdot {y_{t-1}^{1}}^{0{,}24} \cdot {X_t^{1}}^{1{,}997} \cdot {X_t^{2}}^{-2{,}867} \cdot {X_t^{3}}^{1{,}163} \cdot {X_t^{4}}^{0{,}427} \cdot {X_t^{5}}^{-0{,}321} \cdot {y_t^{3}}^{-0{,}380} \cdot {y_t^{5}}^{-0{,}185} \cdot {y_t^{7}}^{0{,}085}$$ [TB2](8)

6. Conclusions

For estimating the function parameters, the linear Ordinary Least Squares (OLS) method is used. OLS for the following model is:

$$A_t^1 = \alpha_0 + \alpha_1 A_{t-1}^1 + \beta_1 B_t^1 + \beta_2 B_t^2 + \beta_3 B_t^3 + \beta_4 B_t^4 + \beta_5 B_t^5 + \delta_3 C_t^3 + \delta_5 C_t^5 + \delta_7 C_t^7$$

where: e - the natural logarithm figure $Y_i (i = 1,3,5,7); X_j (j = 1,2,3,4,5)$.

$$A_1 = -30.9355 + 0.24\, A_{1_{t-1}} - 0.38026\, A_{3_t} - 0.18513\, A_{5_t} + 0.08539\, A_{7_t} +$$
$$1.99797\, B_{1_t} - 2.86778\, B_{2_t} + 1.16342\, B_{3_t} + 0.42739\, B_{4_t} - 0.32168\, B_{5_t}$$

The α_i - coefficients estimate the elasticity of environmental shocks from exogenous factors. The coefficient of elasticity of the factor shows the percentage of the function increase if the factor increases by 1%.

When $\sum \alpha_i > 1$ the function grows faster than the average, and at $\sum \alpha_i < 1$ it grows slower. In our case, $\sum \alpha_i > 1$. When $\sum \alpha_i > 1$, the average costs per unit of environmental shocks decreases with the increasing number of environmental shocks.

The cumulative impact of the considered exogenous factors for the analysed period increased, however, with a decrease in average costs, because the sum of the elasticity coefficients of the production function is more than 1.

The coefficient α_0 = - 30.935 shows that the exogenous factors had a neutral impact.

The indicator "environmental shocks" in the previous period has a positive impact on itself in the current period. It shows that the risk of shocks is increasing, which is the result of anthropogenic and natural forces.

The indicators "greenhouse gas emissions from industry" and "global temperature change" compared to the other indicators have less influence on the indicator "environmental shocks". Despite this, with the increase in greenhouse gas emissions and global temperature change, the whole ecological system changes and environmental degradation inevitably leads to the development of natural disasters. Global temperature change contributes to the increasing number of the climate-related disasters – raising the level of the ocean and flooding coastal areas. The indicator "area of forested land" has a positive effect, demonstrating that the increase in forested land reduces the risk of climate disasters due to the absorption of greenhouse gas emissions from industry. The indicator "global GDP per capita" reflects a positive effect, i.e. by increasing the production of goods and services in the world the number of negative effects and the risk of technological disasters increases. Rail transport freight turnover and mining of coal and lignite as well as extraction of peat forces us to spend more energy, which in turn increases the release of harmful substances into the atmosphere. The indices "global merchandise exports" and "economically active global population" demonstrate a positive impact in this analysis, i.e. an increase in the export of goods due to transportation also increases the emission of harmful gases into the atmosphere.

Acknowledgements
This paper is based on research carried out with the financial support of the grant of the Russian Science Foundation (Project No. 14-38-00009). Peter the Great St. Petersburg Polytechnic University.

References

[1] Loftness V and Snyder M 2012 Sustainable and Healthy Built Environment *Encyclopedia of Sustainability Science and Technology* 10260–10285
[2] Mohapatra S, Adamowicz W and Boxall P 2016 Dynamic technique and scale effects of economic growth on the environment *Energy Economics* **57** 256–264
[3] Chen Y and Shen T-J 2017 A general framework for predicting delayed responses of ecological communities to habitat loss *National Center for Biotechnology Information* **7 (1)**
[4] Klijn F 1994 Ecosystem Classification for Environmental Management *Ecology & Environment* **2** 49-67
[5] Rudman S M et al 2017 Evosystem Services: Rapid Evolution and the Provision of Ecosystem

Services *Trends in Ecology & Evolution* **32 (6)** 403-415

[6] Klomp J 2016 Economic development and natural disasters: A satellite data analysis *Global Environmental Change* **36** 67-88

[7] Agostino G, J Dunne J P and Pieroni L 2016 Government Spending, Corruption and Economic Growth *World Development* **84** 190-205

[8] Wesseh P K and Lin B 2016 Modelling environmental policy with and without abatement substitution: A tradeoff between economics and environment? *Applied Energy* **167** p 34-43

[9] Scharin H et al 2016 Processes for the sustainable stewardship of marine environments *Ecological Economics* **128** 55-67

[10] Marin G and Modica M 2017 Socio-economic exposure to natural disasters *Environmental Impact Assessment Review* **64** 57-66

[11] Didenko N, Rudenko D and Skripnuk D 2015 Environmental security issues in the Russian Arctic. *Int.l Mult. Sc. GeoConf. Surveying Geology and Mining Ecology Management* **3 (5)** 267-274

[12] Didenko N I, Skripnuk D F and Mirolyubova O V 2017 Urbanisation and Greenhouse Gas Emissions from Industry *IOP Conf. Series: Earth and Environmental Science* **72** 012014

[13] Joseph C et al 2017 Use of significance thresholds to integrate cumulative effects into project-level socio-economic impact assessment in Canada *Environmental Impact Assessment Review* **67** 1-9

[14] Ciffroy P, Péry A.R.R and Roth N 2016 Perspectives for integrating human and environmental exposure assessments *Science of The Total Environment* **568** 512-521

[15] Cardenas I C and Halman J.I.M 2016 Coping with uncertainty in environmental impact assessments: Open techniques *Environmental Impact Assessment Review* **60** 24-39

[16] Bhatt D and Mall R K 2015 Surface Water Resources, Climate Change and Simulation Modelling *Aquatic Procedia* **4** 730-738

[17] Vyboldina E et al 2016 Analysis of Export Restrictions and their Impact on Metals World Markets *Indian Journal of Science and Technology* **9 (5)**

[18] Johansson L and Alderton W D 2015 Mineral Resources *Reference Module in Earth Systems and Environmental Sciences* Available from: https://www.elsevier.com/solutions/sciencedirect/content/reference-modules/earth-system [Accessed 10th May 2018]

[19] Pamukcu P et al 2016 Ecohydrologic modelling of water resources and land use for watershed conservation *Ecological Informatics* **36** 31-41

[20] Ghosh A et al 2015 Theoretical Population Biology **100** 1-5

[21] Hancock T et al 2017 One planet regions: planetary health at the local level *The Lancet Planetary Health* **1 (3)** e92-e93

[22] Romashkina G F, Didenko N I and Skripnuk D F 2017 Socioeconomic modernisation of Russia and its Arctic regions *Studies on Russian Economic Development* **28 (1)** 22-30

[23] Cord A F et al 2017 Priorities to Advance Monitoring of Ecosystem Services Using Earth Observation *Trends in Ecology & Evolution* **32 (6)** 416-428

[24] Mirolyubova O et al 2017 Analysis of world greenhouse gas emissions caused by agreeculture *Int. Mult. Sc. Conf. on Social Sciences and Arts SGEM 2017* **IV** 57-64

Assessment of the Influence of the Extraction of Energy Resources on the Environment

N I Didenko, E S Romashkina

Peter the Great St. Petersburg Polytechnic University, St. Petersburg, 195251 Russia

didenko.nikolay@mail.ru

Abstract. In the paper a proposed model for evaluating the influence of energy resource extraction activities on indicators for assessing the state of the environment is considered. The model was constructed under several different assumptions. The first suggests that there exists an environment in which different processes are taking place, e.g. forests and fresh water stocks are changing along with temperature. These processes are to be assessed by endogenous variables. Moreover, each process has its own prehistory that is also being affected by each endogenous variable. The second assumption states that all the processes taking place in the environment are interrelated. The third assumption supposes that the environment is affected by external factors, primarily human activity. In the current paper the human activity of energy resource extraction was considered. The model comprised four econometric equations, each of which was considered to be an ADL model. The model represented the coefficient calculation methodology with time series stationarity and multicollinearity verification. Identification of endogenous variable lags correlated strongly with the current period variable value and verification of the significance of autocorrelation coefficients. The findings of the study allowed conclusions to be drawn, which offered directions for further investigation.

1. Introduction

The state of the environment, represented in terms of processes occurring in it [1, 2], is in a constant state of flux. Among these processes, scientists have distinguished such phenomena as ozone depletion, climate change, reduced freshwater supply, enlargement of oceanic dead zones and many others [3, 4]. When estimating these processes in figures, it can be stated that over the 24-year period from 1992 to 2016 the level of CO_2 emissions has increased by 62.1%; the temperate level raised by 167.6%; freshwater resources per capita (1000 m^3) reduced by 26.1%; oceanic dead zones indicator (number of affected regions) increased by 75.3% [5].

All these adverse tendencies appear to be due to different external factors affecting the state of the environment. Among other anthropogenic (human impact) factors, the activity of extracting energy resources can be distinguished. A rapid increase in the human population of the Earth has led to the extraction of more and more raw materials to satisfy their needs [6] and resulted in an increased burden on nature beyond reasonable limits. It should be taken into account that the state of the environment is being influenced by both external factors, including anthropogenic, as well as by other processes occurring within the environment itself. This is because all processes occurring in the environment must be considered to be interrelated [7, 8]. This assumption underpins the definition of the aim of the current survey. This aim is focused on assessing the environment effect of the human activity of energy resource extraction in the context of other processes occurring within the environment itself.

2. Data

The data were collected using the National Aeronautics and Space Administration (hereinafter, NASA) [9], National Oceanic and Atmospheric Administration (hereinafter, NOAA) [10], United Nations Environment Program (hereinafter, UNEP) [11], World Energy Yearbook [12] and World Resources Institute databases [13]. The research covered the period from 2000 to 2016. The global volume of CO2 emissions indicator (including Land-Use Change and Forestry) was calculated as the global volume of CO2 emissions produced by all the countries of the world in million tonne of carbon dioxide per year. The global mean sea level indicator contained the global mean sea level generated from the Integrated Multi-Mission Ocean Altimeter Data for Climate Research (hereinafter, GMSL dataset). The GMSL was considered to be a 1-dimensional time series of globally averaged Sea Surface Height Anomalies (hereinafter, SSHA) from TOPEX/Poseidon, Jason-1 and OSTM/Jason-2 [9]. This has been in operation from September 1992 to the present, with a lag of up to 4 months. Data was reported as changes relative to January 1, 1993 and presented in millimetres. All biases and cross-calibrations applied to the data so SSHA were consistent between satellites. Glacial Isostatic Adjustment (GIA) was applied [14].

The global desertification affected area indicator was calculated in accordance with United Nations standards. For the purposes of this indicator, desertification was defined as land degradation in arid, semi-arid, and dry sub-humid areas resulting from various factors, including climatic variations and human activities (UN Convention to Combat Desertification, 1994). In contradistinction with polar and sub-polar regions, arid, semi-arid, and dry sub-humid areas were considered as areas in which the ratio of annual precipitation to potential evapotranspiration falls within the range from 0.05 to 0.65 (UN Convention to Combat Desertification, 1994). The global desertification affected area indicator was measured in million hectares.

The global land and oceanic temperature anomalies indicator was calculated in °C with respect to the 20th century average. The global oil extraction volume indicator was defined as the global volume of oil extracted by all the countries of the world in million tonnes per year. The global natural gas volume extraction indicator included the global volume of gas extracted by all the countries of the world in billions of cubic metres per year. The global volume of coal extraction indicator covered the global volume coal extracted by all the countries of the world in million tonnes per year.

3. Methodology

The methodology was subdivided into several operational stages: selection of variables assessing the environment state and factors influencing the state the environment; formulation of assumptions for creating the model; selection of appropriate model meeting the assumptions made; calculation of coefficients of the model equations: time series stationarity verification; multicollinearity verification; identification of endogenous variable lags having a strong correlation with the current period variable value and verification of the autocorrelation coefficients significance using Ljung-Box Q-test.

3.1. Selection of variables assessing the environment state and factors influencing the state of the environment

The state of the environment is represented in processes occurring in it such as climate change, freshwater stock reduction and deforestation. Thus, the environment state can be assessed with variables measuring these processes. Endogenous variables are as follows:

y_t^1 - global volume of CO2 emissions, in million tonnes of carbon dioxide per year [13];
y_t^2 - global mean sea level change, in millimetre per year as changes relative to January 1, 1993 [9];
y_t^3 - global total desertification affected areas, in million hectares [11];
y_t^4 - global land and ocean temperature anomalies with respect to the 20th century average, in °C [10].

The state of the environment is being constantly affected by different factors. The energy resources extracting activity factor was distinguished among them as most severe one. The energy resources extracting activity was assessed with the following exogenous variables:

x_t^1- global volume of oil extraction, in million tonnes per year;

x_t^2 - global volume of natural gas extraction, in billions of cubic meters per year;
x_t^3 - global volume of coal extraction, in million tonnes per year.

The values of the variables were presented in Table 1. The descriptive statistics of variables are shown in Table 2.

Table 1. Environmental assessment variables and factors

Year	y_t^1 Global CO2 emissions in million tonnes of carbon dioxide per year	y_t^2 Global mean sea level change in millimetres per year as changes relative to January 1, 1993	y_t^3 Global total desertification affected areas in millions of hectares	y_t^4 Global land and ocean temperature anomalies relative to the 20th century average, in °C	x_t^1 Global volume of oil extraction in millions of tonnes per year	x_t^2 Global volume of natural gas extraction in billions of cubic metres per year	x_t^3 Global volume of coal extraction in millions of tonnes per year
2000	27560	22.1	3510	0.42	3597	2509	4601
2001	28800	24.4	3517	0.54	3626	2536	4822
2002	29230	28.2	3524	0.60	3636	2617	4879
2003	29800	32.9	3531	0.61	3715	2705	5225
2004	30980	35.7	3538	0.58	3581	2792	5610
2005	31800	38.8	3545	0.65	3860	2863	5969
2006	33750	40.9	3550	0.61	3824	2931	6308
2007	32100	43.5	3556	0.61	3903	3057	6573
2008	33560	43.2	3561	0.54	3990	3152	6768
2009	32160	47.6	3567	0.63	3244	3072	6909
2010	34570	53.8	3572	0.70	3956	3324	7296
2011	35120	49.1	3577	0.58	3988	3388	7726
2012	36100	57.9	3584	0.62	4022	3452	7830
2013	36810	67.3	3590	0.67	4038	3525	7865
2014	37100	66.0	3595	0.74	4102	3278	7847
2015	39700	73.0	3600	0.91	4185	3563	7648
2016	41000	82.8	3607	0.95	4227	3629	7168

Source: data published by the National Aeronautics and Space Administration, National Oceanic and Atmospheric Administration, United Nations Environment Program, World Energy Yearbook, World Resources Institute databases [9], [10], [11], [12], [13], [15].

Table 2. Descriptive statistics for environmental assessment variables and factors

Variable	Meaning of variable	Measured in	Minimum value	Maximum Value	Mean	Standard deviation
y_t^1	Global CO2 emissions	in million tonnes of carbon dioxide per year	27560	41000	33538	3804
y_t^2	Global mean sea level change	in millimetres per year as changes relative to January 1, 1993	22	82	47	17
y_t^3	Global total desertification affected areas	in millions of hectares	3510	3607	3560	30
y_t^4	Global land and ocean temperature anomalies	with respect to the 20th century average, in °C	0.42	0.95	0.64	0.13
x_t^1	Global volume of oil extraction	in millions of tonnes per year	3244	4227	3853	257
x_t^2	Global volume of natural gas extraction	in billions of cubic metres per year	2509	3629	3082	371
x_t^3	Global volume of coal extraction	in millions of tonnes per year	4601	7865	6532	1155

Source: data calculated by the author and initially collected from publications of the National Aeronautics and Space Administration, National Oceanic and Atmospheric Administration, United Nations Environment Program, World Energy Yearbook, World Resources Institute databases [9], [10], [11], [12], [13].

3.2. Formulation of assumptions for creating the model

The first assumption states that there is an environment in which different processes take place: forest and fresh water stocks are varying alongside changes in temperature. All of these processes are assessed using endogenous variables and have their own prehistory by which they are affected. The second assumption is that all processes taking place in the environment are interrelated; for example, changing forestation affects the level of CO_2 emissions. The third assumption presupposes that the environment is affected by external factors including human activity. In the present paper, the human activity or anthropogenic factor of energy resource extraction is considered. In short form, all the assumptions made can be introduced in the following way:

$$y_t^1 = \varphi(y_{t-j}^1; y_t^2; y_t^3; y_t^4; x_t^1; x_t^2; x_t^3) \quad (1)$$
$$y_t^2 = \varphi(y_{t-j}^2; y_t^1; y_t^3; y_t^4; x_t^1; x_t^2; x_t^3) \quad (2)$$
$$y_t^3 = \varphi(y_{t-j}^3; y_t^1; y_t^2; y_t^4; x_t^1; x_t^2; x_t^3) \quad (3)$$
$$y_t^4 = \varphi(y_{t-j}^4; y_t^1; y_t^2; y_t^3; x_t^1; x_t^2; x_t^3) \quad (4)$$

3.3. Selection of appropriate model for the assumptions made

In the previous stage we supposed that there were dependences between processes occurring in the environment. These processes could be described through different types of equations. The question was what types of equations demonstrated these dependences. In order to model the assumptions, the system of econometric equations, each of which was presented in ADL model form, was considered to be the most suitable since the endogenous variables depended on both the other periods and other time series. In general form, the ADL model was represented as follows:

$$y_t^1 = \alpha_0 + \alpha_1 y_{t-1}^1 +..+\alpha_i y_{t-i}^1 +..b_i y_t^k + c_j x_t^k + \varepsilon \quad (5)$$

where y_{t-i}^1 – value y_t^1 with lag t-i, demonstrating the prehistory of the process;
y_t^k – time series showing kth endogenous variable;
x_t^k - time series showing kth exogenous variable.

3.4. Calculation of the model equation coefficients

The methodology of coefficients calculation included several steps: time series stationarity verification; multicollinearity test for endogenous and exogenous variables; identification of the endogenous variable lags having a strong correlation with the current period variable value and verification of the autocorrelation coefficients significance using the Ljung-Box Q-test.

3.4.1. Time series stationary verification. Having chosen the endogenous and exogenous variables, we checked the time series of variables for stationarity, using Dickey-Fuller test. Dickey-Fuller test supposed finding the coefficient a in the following autoregressive equation:

$$y_t = ay_{t-1} + \varepsilon_t \quad (6)$$

where y_t— time series, and ε_t— error. According to the Dickey-Fuller test, all the time series appeared to be stationary because coefficient $|a| < 1$ for all of them.

3.4.2. Multicollinearity verification. During this step all the endogenous and exogenous variables were tested for multicollinearity. Coefficients of pair correlations were found and analysed. According to the analysis carried out, no variables were excluded from the equations.

3.4.3. Identification of the endogenous variables' lags. During this step, the endogenous variable lags which had a strong correlation with the current period endogenous variable value, were identified. In order to determine these, the autocorrelation coefficients were first calculated. The verification of the significance of the autocorrelation coefficient was made with Ljung-Box Q-test and Box-Pierce Q-statistic. The endogenous variable lags with autocorrelation coefficient significance of more than 0.7 were included in equations. The calculations show that the endogenous variable y_t^1 is dependent on y_{t-7}^1, since the autocorrelation coefficient equals -0.786. The endogenous variable y_t^2 showed no dependence on the values of the other periods. The endogenous variable y_t^3 was dependent on the y_{t-9}^3, since the autocorrelation coefficient appeared to be -0.855. The endogenous variable y_t^4 showed no significant dependence on the values of the other periods. The model coefficients were calculated using the method of least squares. The significance of the regression equations and coefficients of the regression was evaluated. Finally, the model structure, including calculated coefficients, was written down as represented in system of equations 7:

$$\left\{\begin{array}{c} y_t^1 = -337531.042 + 0.007 \times y_{t-7}^1 + 17.085 \times y_t^2 + 100.616 \times y_t^3 + 2429.786 \times y_t^4 + \\ +2.638 \times x_t^1 + 1.939 \times x_t^2 - 0.885 \times x_t^3 \\ \\ y_t^2 = -2527.114 + 7.337E - 0.5 \times y_t^1 + 0.728 \times y_t^3 + 5.631 \times y_t^4 + 0.001 \times x_t^1 + \\ +0.007 \times x_t^2 - 0.008 \times x_t^3 \\ y_t^3 = 3427.524 - 0.001 \times y_{t-9}^3 + 0.001 \times y_t^1 + 0.708 \times y_t^2 + 15.925 \times y_t^4 - 0.004 \times x_t^1 - \\ -0.004 \times x_t^2 + +0.011 \times x_t^3 \\ y_t^4 = -19.936 + 0.00001 \times y_t^1 + 0.003 \times y_t^2 + 0.006 \times y_t^3 - 0.00001 \times x_t^1 \end{array}\right\} \quad (7)$$

The coefficients of determination for the equations varied from 0.893 to 0.994 with F-criterion ranging from 13 to 374 and significance of 0.001 for all equations.

4. Conclusion

The evidence from the study allowed conclusions to be drawn concerning the network structure between variables assessing the environmental state and factors influencing the state of the environment. We report two sets of findings. The first describes the significant interrelations between the variables assessing the environment state and factors influencing this state. The second distinguishes significant interrelations between the variables assessing the processes occurring in the environment.

The first set of findings could be put down as followed. There existed a link between the global volume of oil extracted and the global volume of CO_2 emissions. A one-million-tonne increase in oil extracted led to an increase of the global volume of CO_2 emissions by 2.636 million tonnes of carbon dioxide with probability at p = 0.955. The interrelations between the global volumes of natural gas, coal extracted and the global volume of CO_2 emissions appeared with probability at p = 0.513 and coefficients equalling + 1.939, - 0.885 respectively.

The interrelations between the global volumes of oil and natural gas extraction and the global mean sea level change indicator were considered to be with probability at p = 0.246 and p = 0.558 respectively; the coefficients were equal to + 0.001, + 0.007 respectively. Surprisingly enough, there appeared to be an inverse link between the global volume of coal extraction and the global mean sea level change indicator. A one-million-tonne rise in the amount of coal extracted led to a fall in the global mean sea level change indicator (measured as change relative to January 1, 1993) by – 0.008 millimetres with probability at p = 0.967.

The interrelations between the global volumes of oil, natural gas extraction and expansion of total global desertification affected areas were considered to be with probability at p = 0.552 and p = 0.333 accordingly, with coefficients equal to - 0.004 and - 0.004 respectively. At the same time the model demonstrated the direct link between the global volume of coal extracting and expansion of the global total desertification affected areas. A one-million-tonne increase in coal extracted appeared to be interrelated with a rise in desertification of affected areas by 0.011 million hectares with probability at p = 0.971.

The interrelations between the volumes of oil and natural gas extraction and the global land-ocean temperature anomalies indicator relative to the 20th century average had a probability of p = 0.096 and p = 0.445 respectively.

The second set of findings distinguishing the significant interrelations between the variables assessing the environment themselves may be described as follows. The model demonstrated a direct link between the global volume of CO2 emissions and the expansion of the total global area affected by desertification. An increase in total global desertification with affected areas of 1 million hectares led to increase of the volume of CO_2 emissions by 100.616 million tonnes of carbon dioxide with probability at p = 0.700. It seems that land degradation represented in terms of processes including deforestation caused in its turn an increase in the global volume of CO_2 emissions to the environment.

The model demonstrated a direct link between the global mean sea level change indicator and expansion of the global total area affected by desertification. The expansion of the global total desertification affected areas by 1 million hectares appeared to be interrelated with the rise of global mean sea level change indicator (measured as change relative to January 1, 1993) by 0.728 millimetres with probability at p = 0.94.

The increase in the global land and ocean temperature anomalies change indicator (measured relative to the 20th century average) by 0.01 °C (1 unit was treated as a change of 0.01 °C with respect to the 20th century average) was interrelated with the increase in expanding the global total desertification affected areas by 15.925 million hectares with probability at p = 0.72.

The analysis revealed that for two variables only the prehistory assumptions were correct. It should be mentioned that only significant interrelations were highlighted in conclusions detailed in the current paper. The analysis demonstrated the methodology for assessing the influence of human activity of energy resource extraction on indicators assessing the state of the environment. The methodology could be applied in other spheres under the assumptions made. Having analysed and applied this methodology in environment sphere, the authors found it possible to apply it in their further researches in the sphere of banking sector assessment. In order to obtain results having high significance, it is recommended that a time series with more than thirty years be used. For more accurate outcomes, the number of exogenous variables should exceed the number of endogenous ones by at least three times.

Acknowledgements

The paper is based on research carried out with the financial support of the grant of the Russian Science Foundation (Project No. 14-38-00009, The program-targeted management of the Russian Arctic Zone development). Peter the Great St. Petersburg Polytechnic University.

References

[1] Schlesinger William H et al 1990 Biological Feedbacks in Global Desertification *Science* doi: 10.1126/science.247.4946.1043
[2] Middleton N, Thomas D S G and Edward A 1992 World atlas of desertification (United nations environment programme) doi: 10.1002/ldr.3400030407
[3] Mann M E 2014 Earth Will Cross the Climate Danger Threshold by 2036 *Scientific American* Available from: https://www.scientificamerican.com/article/earth-will-cross-the-climate-danger-threshold-by-2036/ [Accessed 10th May 2018]
[4] Ripple W J et al 2014 Ruminants, climate change and climate policy *Nature Climate Change* doi:10.1038/nclimate2081
[5] Ripple W J et al 2017 World Scientists' Warning to Humanity: A Second Notice *BioScience* doi:10.1093/biosci/bix125
[6] Didenko N I, Skripnuk D F and Mirolyubova O V 2017 Urbanization and Greenhouse Gas Emissions from Industry *IOP Conf. Series: Earth and Environmental Science* **72** 012014
[7] Bowman T E et al 2010 Time to take action on climate communication *Science* doi: 10.1126/science.330.6007.10
[8] Didenko N I et al 2017 Deforestation and human activity: A global perspective *International Mult. Sc. GeoConf. Surveying Geology and Mining Ecology Management* **17 (53)** 183-190
[9] NASA National Aeronautics and Space Administration Global mean sea level Global time series Available from: https://sealevel.nasa.gov [Accessed 27th May 2018]
[10] NOAA National oceanic and atmospheric administration national centers for environmental information climate at a glance: Global time teries 2018 Available from: https://www.ncdc.noaa.gov/cag/global/time-series [Accessed 27th May 2018]
[11] UNEP United Nations environment program Available from: https://www.unenvironment.org [Accessed 27th May 2018]
[12] World energy yearbook Global time series Available from: https://yearbook.enerdata.ru [Accessed 27th May 2018]
[13] Climate data explorer 2015 Washington DC world resources institute Available from: http://cait.wri.org [Accessed 27th May 2018]
[14] Beckley B D et al 2017 On the cal-mode correction to Topex satellite altimetry and its effect on the global mean sea level Time series doi:10.1002/2017JC013090
[15] Skripnuk D et al 2017 The formation of the regional tourist cluster in the Euro-Arctic Region *Int. Mult. Sc. Conf. on Social Sciences and Arts, SGEM 2017* **IV** 369-374

IOP Conf. Series: Earth and Environmental Science **180** (2018) 012015 doi:10.1088/1755-1315/180/1/012015

Territorial-sectoral modelling of the automotive industry in the Russian Federation

K Kikkas

Peter the Great St. Petersburg Polytechnic University, St. Petersburg, 195251 Russia

xekikkas@gmail.com

Abstract. The paper describes a territorial-sectoral model for the development of the automotive industry. This model is based on a novel institutional-functional approach proposed by author. It explains the choice of the indicators assessing the main areas of human activity on the territory and the indicators evaluating the automotive industry that operates on this territory. The distinctive feature of the institutional-functional approach is that it includes both such variables as spheres of human life on the territory in which the automotive industry functions and variables of the automotive industry into the model. The model includes three levels, each of which has the system of equations, from 6 to 8 equations. At each subsequent level, the endogenous indicators of the previous level become exogenous. The model comprises a system of interrelated and interdependent econometric equations. Each equation is formulated as an ADL model. Within the system of interrelated and interdependent econometric equations, a correlation between endogenous and exogenous variables in the automotive industry and spheres of human activity in the territory is demonstrated. A three-level model for the development analysing of the Russian automotive industry is proposed. On the basis of the constructed model, a short-term forecast until 2020 is presented.

1. Introduction

The automotive industry, which encompasses the production of all kinds of machinery, equipment and appliances, is one of the most extensive industries of the Russian Federation. Currently in this area there are more than 200 manufacturers. The automotive industry comprises 31% of the total volume of machine-building complex in the world [1]. In developed countries, the car industry typically provides five to ten percent of total gross national production.

The automotive industry is directly related to the products of other industries (non-ferrous metals, electronic products, rubber, and others). The automotive industry is also related to trade and maintenance of vehicles, construction and repair of roads, etc. The car industry is a prime example of monopolistic competition, i.e. a huge number of companies produce differentiated products. Such goods are related but not completely interchangeable because they provide brand products different from their competitors.

The car industry is divided on the following grounds: the condition of the car (the market for new and second-hand cars); geographical (markets in USA, Japan, China, Russia and other regions); vehicle type (cars and trucks, special vehicles, buses, ATVs and motorcycles) [2].

According to Forbes magazine, 2018 is the third consecutive year when the world's three largest automakers remain the same: Toyota Motor, Volkswagen Group and Daimler. The Russian company JSC "AVTOVAZ" entered the list of the reputable economic magazine 'Forbes' only in the middle of the two thousandths: in 2005 it took the 1836th place and in 2006 it rose to the 1653th place [http://www.oica.net]. Since the automotive industry is widely seen as one of the vital economic sectors for economic growth and social development, industry production is widely used as an economic metric. The automotive industry is developing towards modernisation of designs, reliability improvement, fuel economy improvement, reduced emissions and use of new materials as well as improvements in the computerisation of maintenance, management and security.

One of the tools for analysing the activities of the automotive industry is modelling [4, 5]. The purpose of the present study is to design a mathematical model for the development of the Russian automotive industry. The mathematical model should contain several interdependent equations. Each equation connects endogenous and exogenous variables. Such a model should be used both for analysis and for forecasting the growth indices of the automotive industry.

The local tasks of the study are the following: describe the object of research (the automotive industry); choose the type of mathematical model; describe the method of empirical verification of the model; check the model using the statistics of the automotive industry for 2000-2015.

2. Review of the problem research

There are many approaches used in the modelling of some object, including an industry or a territory. This is confirmed by the studies of M. Porter, David P. Byrne [6], Lucija Muehlenbachs [7], G. F. Romashkina [8], N. I. Didenko [9], D. F. Skripnuk [10], Murray Z. Frank [11], Jennifer F. Reinganum [12] and other scientists.

3. Approach to territorial and sectoral modelling

3.1. Institutional-functional approach to territorial-sectoral modelling

A distinctive feature of the institutional-functional modelling approach, on which the territorial-sectoral industrial development model is based, is that it includes both such variables as spheres of human life on the territory in which the automotive industry functions as well as variables of the automotive industry itself [13-16]. The institutional component is provided by an analysis of human activity spheres on the territory, while the functional component consists in an analysis of the automotive industry functioning in the area. This approach permits important data to be obtained, informing the future direction of development of both the territory and the industry.

3.2. The theoretical model and its characteristics

The model includes three levels, each of which comprises a system of between 6 and 8 equations. The model of the first level displays the production environment, i.e. the territory. In developing the model at the first level, each enterprise or industry is expressed through the manufacture function which has the following form: $Q = f(A, L, K)$, where Q is the volume of shipped products of the enterprise/industry, A is the technology index (innovational development of the enterprise/industry), L is the number of employees, K is the physical capital (value of fixed assets).

The model of the second level displays the spheres of human activity taking place in the area (demographic, social, productive and economic, environmental, innovative-technological, political and spiritual). The model of the third level displays the correlation of the territorial-space industry with the national economy. In the model of each next level, the endogenous indicators of the previous level become exogenous.

A detailed model of the third level is presented below. As endogenous indicators, the following six variables were selected:

y_t^{31} – proportion of export of cars from Russia to the total exports of the Russian Federation;

y_t^{32} – proportion of export of lorries from the Russian Federation in the total exports of the Russian Federation;

y_t^{33} – proportion of importation of cars from the Russian Federation in the total imports of the Russian Federation;
y_t^{34} – proportion of importation of lorries from the Russian Federation in the total imports of the Russian Federation;
y_t^{35} – proportion of the automotive industry relative to Russia's GDP;
y_t^{36} – the specific weight of production volume of passenger cars in the total production of vehicles in Russia.
The exogenous variables for endogenous variable y_t^{31} are:
x_t^{231} – volume of exports in the Russian Federation (in mln. US dollars);
x_t^{311} – volume of passenger cars' export (in mln. US dollars);
x_t^{312} – index of labour productivity in the automotive industry in Russia (%);
x_t^{313} – consumer price index for goods and services of the automotive industry in Russia (in %).
The exogenous variables for endogenous variable y_t^{32} are:
y_t^{1} – volume of vehicle production (units);
x_t^{231} – volume of the Russian Federation exports (mln.US dollars);
x_t^{321} – export volume of trucks (mln.US dollars);
x_t^{322} – share of investments into the automotive industry relative to the total volume of investments in the fixed capital of the Russian Federation (in %).
The exogenous variables for endogenous variable y_t^{33} are:
x_t^{232} – volume of the Russian Federation imports (mln. US dollars);
x_t^{331} – volume of passenger cars' imports (mln. US dollars);
x_t^{212} – level of motorisation (cars/1,000 people);
x_t^{313} – consumer price index for goods and services in the automotive industry in Russia (in %).
The exogenous variables for endogenous variable y_t^{34} are:
x_t^{232} – volume of Russian imports (in mln. US dollars);
x_t^{341} – volume of trucks' import (in mln. US dollars);
x_t^{212} – level of motorisation (cars/1,000 people);
x_t^{313} – consumer price index for goods and services in the automotive industry in Russia (in %).
The exogenous variables for endogenous variable y_t^{35} are:
y_t^{2} – contribution of the automotive industry to Russian GDP (in mln. US dollars);
x_t^{351} - GDP in Russia (in mln. US dollars);
x_t^{322} – share of investment into the automotive industry in the total volume of investments into the fixed capital of the Russian Federation (in %);
x_t^{352} - innovative activity of the organisations which produce vehicles on the whole in Russia (in %);
x_t^{252} – Scientific research expenditure (thous. US dollars).

The exogenous variables for endogenous variable y_t^{36} are:

y_t^1 – volume of vehicles' production (units);

y_t^{11} – volume of cars' production in Russia (in units);

x_t^{251} – number of patents (in thousand units);

x_t^{211} – average number of qualified personnel per year (in thousands of people);

x_t^{254} – number of innovation centres in Russia (in units).

The auto regressive distributed lags (ADL) model was selected as a specific type of dependency for each of the six equations.

$$y_t = a_0 + \sum_{i=1}^{n} a_i y_{t-i} + \sum_{j=0}^{q1} b_j x_{t-i}^1 + \ldots + \sum_{j=0}^{qk} b_j x_{t-i}^k + E_t \quad (1)$$

where k is the quantity of exogenous variables;

q_i – the number of lags of the *i-th* exogenous variable $(i = 1,2,\ldots k)$;

n – the depth of the lag within the endogenous variable;

E_t - residues forming the white noise process.

3.3. Data

The informational-operational base comprised data supplied by the Russian federal state statistics service [http://www.gks.ru/Rosstat]; Autostat [https://www.autostat.ru/tags/3/], the statistics of the international organisation of motor vehicle manufacturers (OICA) [http://www.oica.net/]; annual reports of the largest foreign and Russian automotive manufacturers [http://www.honda.co.ru/], [https://www.toyota.ru], [http://ford-rolf.ru], [http://www.lincoln.com], as well as legislative and regulatory acts of the Russian Federation.

3.4. Empirical model testing methodology

The technique consists of the following stages:

a) Collection of statistical data for years 2000 – 2015.

b) Presentation of the indicators in a comparable form.

To bring the indicators into a comparable form, the indexes of each indicator are defined according to the following formula:

$$I_i = \frac{x_i - x_i^{\min}}{x_i^{\max} - x_i^{\min}} \cdot 100 \quad (2),$$

where x_i is the actual value of the i-th index, $x_i^{\max}$ and $x_i^{\min}$ – are maximum and minimum rate of the i-th index, respectively.

c) Inspection of the time series' stationarity using Dickie Fuller test.

d) Implementation of the procedures when composing each of the six equations: analysis of correlation between the endogenous variable of the last period and the exogenous variables; checking statistical multicollinearity; testing the significance of each equation using Fisher's F-criterion and the coefficient of determination; finding the equation coefficients using regression analysis.

e) The view of the model as a system of equations in a structural form. Empirical testing of the system of interrelated econometric equations.

f) Definition of identification terms for each equation.

g) Finding the coefficients of regression equations.

h) Performing predictive calculations using the system of equations.

4. Empirical testing of the model

Statistical data is collected for years 2000-2015 for each indicator. The results are presented in a comparable form. The stationarity of the time series is checked. The stationarity testing revealed that all-time series are stationary, as when finding the coefficients of a regression equation coefficient a is less than the unit for all-time series if the condition $t_{расч.} > t_{табл.}$, $p = 0.05$ is met.

An ADL model is created for each of the six equations. The outcome of the applied techniques was the creation of a system of linked econometric equations in a structural form.

Testing the compliance of the model structural form according to the conditions of identification showed that all the equations are over identified. This procedure was carried out for each equation separately. Therefore, in order to estimate the parameters of each of the equations, a two-step least-squares method was used.

The system of econometric equations with the discovered coefficients of the regression equations is written as:

$$\begin{cases} y_t^{31} = -2.431 + 0.359 \cdot y_{t-1}^{31} + 0.118 \cdot x_{t-2}^{231} + 0.2 \cdot x_{t-1}^{311} + 0.018 \cdot x_{t-1}^{312} + 0.125 \cdot x_{t-1}^{313} \\ y_t^{32} = -31.621 + 0.619 \cdot y_{t-1}^{32} - 0.243 \cdot y_{t-2}^{1} + 0.433 \cdot x_{t-1}^{231} + 0.127 \cdot x_{t-1}^{321} + 0.508 \cdot x_{t-1}^{322} \\ y_t^{33} = 41.198 + 1.40 \cdot y_{t-1}^{33} + 0.603 \cdot x_{t-1}^{232} - 0.644 \cdot x_{t-1}^{331} - 0.918 \cdot x_{t-1}^{212} - 0.246 \cdot x_{t-1}^{313} \\ y_t^{34} = 22.280 + 1.107 \cdot y_{t-1}^{34} + 0.537 \cdot x_{t-1}^{232} - 0.455 \cdot x_{t-1}^{341} - 0.643 \cdot x_{t-1}^{212} + 0.058 \cdot x_{t-2}^{313} \\ y_t^{35} = 6.343 + 0.531 \cdot y_{t-1}^{35} - 1.03 \cdot y_{t-1}^{2} + 1.014 \cdot x_{t-1}^{351} + 0.791 \cdot x_{t-2}^{322} - 0.339 \cdot x_{t-1}^{352} - 0.21 \cdot x_{t-2}^{252} \\ y_t^{36} = 32.006 - 0.072 \cdot y_{t-1}^{36} - 1.049 \cdot y_{t-1}^{1} + 0.015 \cdot y_{t-1}^{11} + 1.308 \cdot x_{t-2}^{251} - 0.602 \cdot x_{t-2}^{211} + 0.906 \cdot x_{t-1}^{254} \end{cases} \quad (3)$$

The accuracy of the model for prediction was determined by comparing the estimated projections for year 2016 with actual statistics of 2016. The deviation of actual values of the control year from the calculation and the forecast turned out to be insignificant, which proves the high accuracy of the model.

5. Use of the model to forecast automotive industry development in the Russian Federation

On the basis of the constructed system of equations and using derived trend equations, a short-term forecast was created up to year 2020. To do this, the following actions were performed:

a) Forecasting of exogenous indicators for the period of four years (2017-2020) was carried out using trend line equations. Forecast values of exogenous parameters are given in the table below (Table 1). Forecasts of the indicators: "volume of exports in the Russian Federation", " volume of export of passenger cars", "production of vehicles in Russia", "volume of imports into the Russian Federation, “level of motorisation", "contribution of the automotive industry to Russia's GDP", "GDP of Russia", "innovative activity of organisations producing vehicles in Russia", " volume of car production", "number of patents" and "average number of highly qualified personnel" will have a positive trend in the next four years.

"Index of productivity", "consumer prices index for goods and services of the automotive industry in Russia", " volume of export of lorries in the Russian Federation", " share of investment into the automotive industry in the total volume of investments into the fixed capital of the Russian Federation", " volume of imports of passenger cars” and “volume of imports of lorries” will have a declining trend. The indicator "number of innovative centres in Russia" will hardly change over the coming years.

According to the forecast results, it can be concluded that the government needs to implement a number of activities, including: improving the quality and image of the Russian automotive industry, technological development in the field, training of employees producing motor vehicles, increase of investment into the automotive industry, expansion of innovation institutions, as well as updating vehicle fleets in the country.

Table 1. Forecasted values for exogenous indicators

Years	x^{231}	x^{311}	x^{312}	x^{313}	y^{1}	x^{321}	
2017	71.976	76.629	16.299	11.842	97.450	33.335	
2018	74.328	80.618	11.365	8.709	101.871	31.664	

2019	76.680	84.606	6.431	5.576	106.291	29.993	
2020	79.032	88.595	1.497	2.443	110.711	28.322	
Years	x^{232}	x^{331}	x^{212}	x^{341}	y^{2}	x^{351}	
2017	95.065	1.892	106.25	19.580	61.735	106.379	
2018	100.781	3.879	112.50	15.102	64.651	112.757	
2019	106.498	9.650	118.75	10.625	67.566	119.136	
2020	112.214	15.421	125	6.148	70.481	125.515	
Years	x^{352}	x^{252}	y^{11}	x^{251}	x^{211}	x^{254}	x^{322}
2017	6.195	102.522	85.2	100.402	87.544	18.294	11.280
2018	0.625	105.043	89	106.695	92.716	18.388	6.931
2019	4.944	107.564	92.7	112.988	97.887	18.482	2.581
2020	10.514	110.085	96.4	119.280	103.059	18.575	1.768

b) The obtained projected values of the exogenous parameters were used in the system of equations for obtaining the endogenous variables. The predicted values of the endogenous variables in years 2017-2020 are shown in the figures (Figures 1, 2, 3, 4, 5, 6).

According to the results of the forecast, the proportion of export of cars in Russia relative to total exports of the Russian Federation, the proportion of imported cars in Russia relative to total Russian imports, the proportion of imported lorries of the Russian Federation to total Russian imports and the proportion of the automotive industry in the total GDP of Russia will gradually grow.

While such indicators as the proportion of exports of lorries in the Russian Federation in total exports of the Russian Federation and the proportion of passenger cars' production volume in the total production of vehicles in the Russian Federation will be reduced.

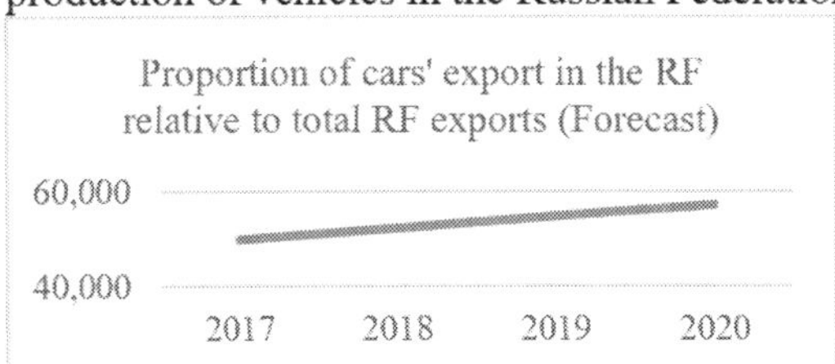

Figure 1. Forecast values for the indicator y_t^{31} **Figure 2**. Forecast values for the indicator y_t^{32}

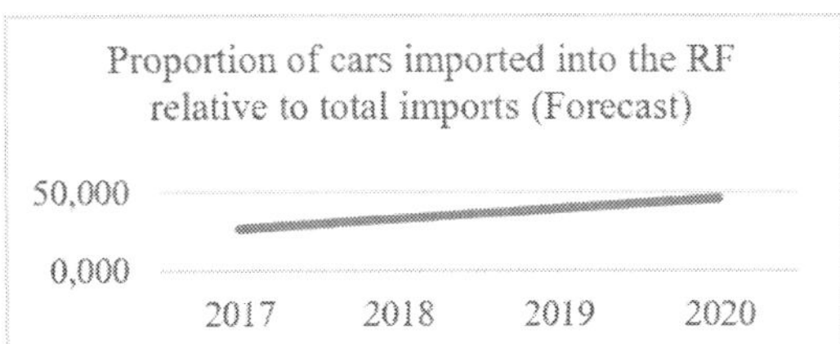

Figure 3. Forecast values for the indicator y_t^{33} **Figure 4.** Forecast values for the indicator y_t^{34}

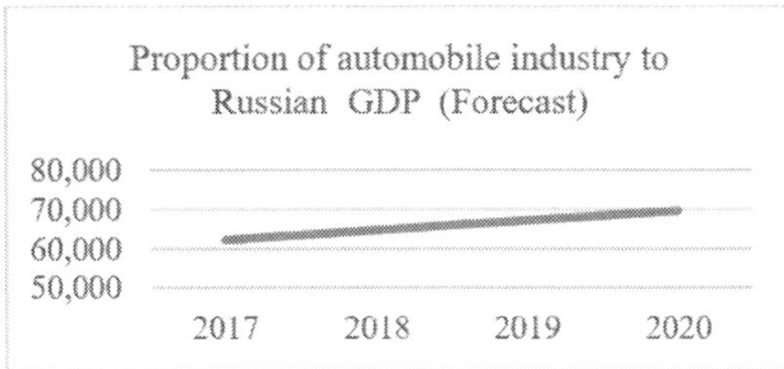

Figure 5. Forecast values for the indicator y_t^{35}

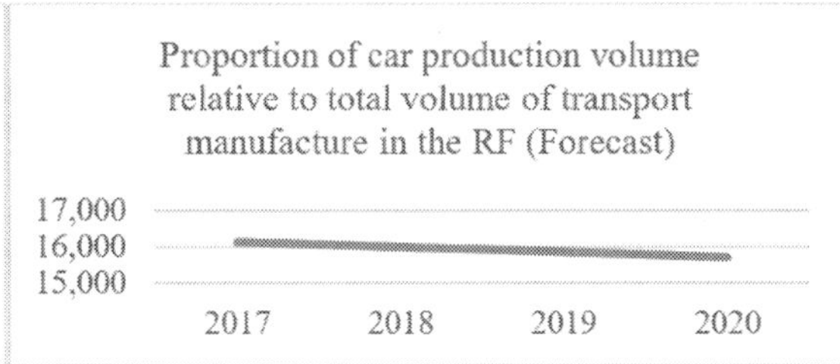

Figure 6. Forecast values for the indicator

6. Conclusion

A three-level model for analysing the development of the Russian automotive industry is proposed. The article analyses the third level of the model. The third level of the model is demonstrated the relationship of the automotive industry with the economy of the country. There are several exogenous variables for each endogenous variable of the model. The endogenous and the exogenous variables selected earlier were used to construct a system of equations in the structural form of the model. The structural form of the model is translated into the reconstructed form of the model. The solution for the reconstructed form of the model is demonstrated. Empirical verification of the model was carried out on the statistical data for 2000-2015. The degree of correspondence between the actual data and those calculated by the model is taken as the criterion for the accuracy of the model. The accuracy of the model was in the range of 5-7%.

Acknowledgements

This paper is based on research carried out with the financial support of the grant of the Russian Science Foundation (Project No. 14-38-00009). Peter the Great St. Petersburg Polytechnic University.

References

[1] Train K E and Winston C 2007 Vehicle choice behavior and the declining market share of U.S. Automakers *International Economic Review* **48 (4)** 1469–1496
[2] Bloom N et al 2013 Does Management Matter? Evidence from India *Q J Econ* **128 (1)** 1-51
[3] Koch T 2016 A Paradigm Shift in the Automotive Industry. *MTZ worldwide* **77 (4)** 74–75
[4] Didenko N and Kunze K 2014 Relationship between Energy Policies and Quality of Life in Russia *WIT Transactions on Ecology and the Environment* **186** 3-11
[5] Rudenko D Y, Pogodaeva T V and Didenko N I 2015 Poverty alleviation strategies in the russian arctic zone regions *Mediterranean Journal of Social Sciences* **6 (1)** 32 -39
[6] Byrne D P 2015 Testing models of differentiated products markets: consolidation in the cable TV industry *International Economic Review* **56 (3)** 805–850
[7] Muehlenbachs L 2015 A dynamic model of cleanup: estimating sunk costs in oil and gas production *International Economic Review* **56 (1)** 155–185
[8] Romashkina G F, Didenko N I and Skripnuk D F 2017 Socioeconomic modernisation of Russia and its Arctic regions *Studies on Russian Economic Development* **28 (1)** 22-30
[9] Didenko N I, Skripnuk D F and Mirolyubova O V 2017 Urbanisation and Greenhouse Gas Emissions from Industry *IOP Conf. Series: Earth and Environmental Science* **72** 012014
[10] Rudenko D and Skripnuk D 2016 Environmental Kuznets curve: The case of arctic Russian regions *Int. Mult. Sc. GeoConf. Surveying Geology and Mining Ecology Management, SGEM2016* **3** 209-216
[11] Frank M Z 1988 An Intertemporal Model of Industrial Exit *Q J Econ* **103 (2)** 333-344
[12] Reinganum J F 1983 Uncertain Innovation and the Persistence of Monopoly *American Economic Review* 73 741-748
[13] Fujiwara-Greve T, Greve H R and Jonsson S 2016 Asymmetry of customer loss and recovery under endogenous partnerships: theory and evidence *International Economic Review* **57 (1)**

3–30
[14] Atroshenko S A, Korolyov I A and Didenko N 2016 Evaluation of physico-mechanical properties of high-chromium tool steels modified with Harrington method *Materials Physics and Mechanics* **26 (1)** 26-29
[15] Mirolyubova O et al 2017 Analysis of world greenhouse gas emissions caused by agreeculture *Int. Mult. Sc. Conf. on Social Sciences and Arts,SGEM 2017* **IV** 57-64
[16] Skripnuk D et al 2017 The formation of the regional tourist cluster in the Euro-Arctic Region *Int. Mult. Sc. Conf. on Social Sciences and Arts, SGEM 2017* 369-374

IOP Conf. Series: Earth and Environmental Science **180** (2018) 012016 doi:10.1088/1755-1315/180/1/012016

Potential Opportunities for the Arctic Transport Space

K Kikkas, E Romashkina

Peter the Great St. Petersburg Polytechnic University, St. Petersburg, 195251 Russia

xekikkas@gmail.com

Abstract. Two distinct routes from Europe to South-East Asia via the seas of the Arctic Ocean are considered in the work: the Northern Sea Route (NSR) and the Northwest Passage. Potential opportunities for the Arctic transport space under the conditions of melting polar ice and expanding navigation period on the Arctic Ocean are substantially increasing. Some alternatives to the NSR and Northwest Passage are described: a) by sea – this is the transport artery going along the Suez Canal; b) by land – these are the transport arteries including the Trans-Siberian Railway (Trans-Sib). The research is aimed at creating a model of the international transport corridor. The model is based on the autoregressive distributed lags (ADL) model. Separate models are constructed for the Northern Corridor, the Trans-Siberian railway, the transport corridor of the Suez Canal and the Northwest Passage. Factors influencing the endogenous variable of the model, consisting in the volume of goods transported for all corridors, are analysed. The exogenous variables were selected separately for each model. The time series covers the 1990-2013 period. The major steps of the implementation of the ADL-model are described: testing the endogenous and exogenous variables for auto-correlation; checking for stationarity of the time series; regression analysis. Finally, conclusions pertaining to the comparison of the NSR with other transport corridors are drawn.

1. Introduction

Although the transport capacity of the Arctic Region is attracting increased attention worldwide, it is of particular interest to the countries of the Arctic region. Arctic routes are of interest due to the possibility to use them to deliver goods from Europe to Asia via the shortest sea route thereby shortening the trip from Asia to America at the expense of cross-polar routes [1]. Strategic interest in the development of the Arctic region is also due to its huge hydrocarbon reserves. Thus, industrial development of the Arctic focusing on the exploitation of hydrocarbon and biological resources will go hand in hand with the development of transport infrastructure and traffic flows in the Arctic Region.

The development of transport operations in the Arctic will require international cooperation in the transport sector, compliance with safety regulations of the areas, harmonisation of legal legislations of the Arctic states and the development of next generation technologies to minimise the impact on the fragile ecosystem. Effective use of the Northern Sea Route (NSR), as well as associated trans-Arctic air corridors, will only be possible if there is a modern infrastructure and full communication and navigation systems.

Several studies have analysed the effects of the economic potential of the transportation in the Arctic zone. In studies [2-4] the forms of international cooperation in the transport sector are presented taking into account the technological complexity of transport operations in the Arctic climate. The legal practices of the Arctic states in field of this cooperation are analysed. In the paper [2] the sea transportation models and their transformation over the last decade are studied and some routes through sea passages in the Canadian Arctic Archipelago are described in detail. The network structure of the various product groups shows the diversity of trade models in the Arctic water between Canadian and foreign fleets. The key challenges for transpolar transportation through the North West Passage are described in terms of climate change, market development and new regulations in [5-7].

In [3] some of the economic risks of exploiting the Arctic sea routes are investigated. Along the coast of Siberia, the average Arctic summer sea ice level has decreased, reducing the time of shipping through the NSR along the Siberian coast from 20 days in the 1990s to 11 days in the years 2012 and 2013. The state of the Arctic Ocean is characterised by a thinning of sea ice and increase of maritime activity. In order to explore the navigation of the Arctic sea routes, a detailed design of oceanic and sea ice (RCP8.5 IPCC emission scenario) is used.

An econometric structural equation model is presented in [4] to test hypotheses explaining how the temperature, rainfalls and wind speed fluctuations affect the passenger and cargo transportation. The model considers changes in traffic rate due to adverse weather conditions in the Arctic region. As a test object, northern Norway's Saltfjellet mountain area is selected. It is proposed to use standardised parameters in transport models to predict the impact of adverse weather conditions on the traffic rate.

The model of the Northern Air Bridge is demonstrated in [8]. The Northern Air Bridge provides air routes from Asia to North America across the Arctic. Mathematical models for analysing the flight routes used by the countries of Southeast Asia are presented. Depending on the flight route, flight times are decreased by 2-5 hours.

The growth of human activity multiplies the problems associated with the management of the Arctic Ocean. In paper [9], a mechanism is proposed for responding to emerging needs in the management of the Arctic Ocean based on the United Nations Convention on the Law of the Sea. This multi-functional mechanism, which is being developed for the regulation of human activities throughout the Arctic Ocean, allows different participants to be assigned to different categories having different rights and responsibilities.

The idea of creating an international Arctic governance framework is not a new one. In paper [10], the development of transport corridors in the Arctic is associated with the exploitation and development dynamics of rich hydrocarbon resources in the Arctic region. When comparing Arctic hydrocarbon exploration and development cycles with global energy prices, unified methods for combating possible pollution of the Arctic shelf are proposed. The oil prices fluctuations, technical problems of oil extraction in the Arctic and issues of sovereignty between circumpolar countries only strengthen the problem of establishing an international framework for the governance of the Arctic.

In study [11, 12], the economic development of Canada's adjacent Arctic shelf is analysed. According to postulates relying on contemporary international law, the country has a significant legal basis for the economic development of the adjacent Arctic shelf. The Northwest Passage is of great importance for Canada. Due to the melting of polar ice, the duration of navigability has increased. In the case of a complete melting of the ice, the strait will be comparable in terms of economic attractiveness to the NSR around the Arctic coast of Russia. The reason for this is that it significantly shortens the route from East Asia to Europe, the United States East Coast and Canada (in comparison with the route through the Panama Canal).

In researches [13], [14], [15], the NSR is compared with the Trans-Siberian transport corridor and the Suez Canal. According to the authors, the Arctic maritime transport route is virtually uncontested in terms of being most effective way of delivering equipment, energy, industrial products and food, all necessary for the functioning of Russian clusters located in the coastal zone of the Arctic seas.

Dodin [16] addresses the issue of education for the sustainable development for the entire circumpolar Arctic, patterns of distribution of important natural resources, the state of the main traffic routes – the NSR and the Northwest Passage – as well as problems of indigenous peoples. For sustainable development in the Arctic, he proposes the establishment of a mechanism in the form of an inter-state government programme dubbed 21st Century Arctic.

Konishev and Sergunin [17] analyse the socio-economic, political, military-strategic and environmental interests of Russia in the Arctic. The authors conclude that these factors are long-term in nature and that the state and society should pay constant attention to them. Expanding the resource base is a priority for Russia in the Arctic zone in order to advance economic development in these areas. At present, the Arctic zone provides 11% of national income in Russia, even though only around 1.4 % of the national population lives there. Mazur [18] considers the Arctic zone as one of the key points of intersection of the various interests in the development of the global world. In the past five years the global importance of the Arctic has increased dramatically in terms of geopolitical and geo-economic processes due to global climate change and the possibility of using natural resources and communications based there. In discussing international cooperation around the Arctic, Tennberg [19] states that associated environmental issues have become a global problem.

In the published Proceedings of the International Conference "The Arctic Region Development and Cooperation", it is noted that the Arctic comprises an ideal region in terms of international cooperation and opportunities for complementary investments.

Since the effective development of the Russian Arctic is impossible without adequate transport provision in the region, the Northern Sea Route (NSR) plays a key role in its development. The NSR is the shortest waterway connecting the western and eastern parts of Russia, as well serving to connect European and Asian ports. In the long run, this transport artery has an excellent chance to become the shortest and most efficient connection between the Asia-Pacific region and Europe.

2. Characteristics of the Northern Maritime Corridors

The Northern Maritime Corridor (NMC) project began in 2002 with the support of the European Union. Its implementation has demonstrated the importance and potential success of cooperation between Russian and Nordic participants. The project participants are representatives of ministries and departments, regional authorities, port authorities and representatives of private companies from eight European countries, including Norway, Russia, Iceland, the Netherlands, Great Britain, and others. The main objective of the project is to create the most favourable conditions for the development of maritime and intermodal transportations, e.g. using different modes of transport through the northern ports of Norway and Russia, as well as to facilitate an integrated transport system in northern Europe. This should help to increase the volume of maritime transport between European ports and the ports of northern Russia. In addition, one of the main driving factors is the development of the oil and gas sector and the implementation of large projects in this area. Using the Northern Maritime Corridor to transport oil and gas equipment is the main goal of the project. At the same time, the establishment of regular posts will attract the attention of forwarding agencies that are already transporting goods to and from Russia through the ports of the Gulf of Finland.

The NSR in Europe begins at Novaya Zemlya Straits (Cape of Desire), and ends in the Bering Strait, Asia. The transportation of goods along the NSR can be carried out along two distinct shipping routes: high-latitude and coastal. The distance along the high-latitude route is 2,200 nautical miles; along the coastal route, it is 2,990 nautical miles. Since the Northwest Sea Passage, Trans-Siberian Railway, and Southern Sea Route are considered to be the most important alternatives to the NSR, these alternatives are incorporated into the econometric analysis. The Northwest Sea Passage comprises a network of several maritime routes through the Canadian Arctic Archipelago, including around 19 thousand islands, as well as many hazardous rocks and reefs. With rapid climate change due to global warming, the shipping lines of the Arctic Ocean are becoming increasingly congested. However, by connecting the Atlantic and Pacific oceans, the Northwest Sea Passage reduces transport routes and thus saves time and money in the implementation of commercial trade. In effect, the melting ice of the Arctic Ocean reduces dependence on the Panama and Suez Canals.

With a total length of 9,288.2 kilometres, the Trans-Siberian Railway, which consists of a railroad network that runs across the Eurasian continent, connecting the European part of Russia comprising its largest industrial areas and capital with its median (Siberia) and eastern (Far East) areas, is the longest railway in the world. In 2002 it became fully electrified.

The key element of the Southern Sea Route is the Suez Canal. The legal status of the Suez Canal is currently regulated by the Convention of Constantinople relating to free passage through the Suez Canal. The most important principles of navigation through the canal, established by the Convention of Constantinople, are the freedom of use of the canal by ships of all countries, the equality of all countries in their use of the canal and the principle of neutrality, implying a prohibition on blockading the Suez Canal.

3. Econometric model of the transport corridors

3.1. Auto regressive distributed lag model

The autoregressive distributed lag (ADL) model was chosen for a formalised description of the economic representation of the transport corridors. ADL is a time-series model, in which the current values depend on the number of past values of the series, as well as on the current and past values of other time series [20, 21].

The ADL model has the form:

$$y_t = a_0 + \sum_{i=1}^{n} a_1 y_{t-i} + \sum_{j=0}^{q_1} b_j x_{t-i}^1 + \ldots + \sum_{j=0}^{qk} b_j x_{t-i}^k + E_t \qquad (1)$$

where k – the number of exogenous variables; q – the number of lags; n – the highest lag, E_t – residues forming white noise process.

3.2. Data

Data on the NSR, the Southern Sea Route, the Trans-Siberian Railway and the Northwest Passage were obtained from the World Bank database, the Administration of NSR Russia, the database of the Research Institute of Arctic and Antarctic regions of Russia, the Ministry of Economic Development and Trade of Russia, the Administration of the Suez Canal, the Administration of the Trans-Siberian Railway, and the Nordic Council.

3.3. Endogenous and exogenous variables used in the model

In all models, the endogenous variable represents the volume of transported goods. The exogenous variables were selected on the basis of their effect on the volume of goods transported and being commonly referred to in the literature on the economics of transportation.

The preliminary analysis considers twenty socioeconomic variables for each corridor (including the aforementioned alternatives). Following preliminary analysis, the number of indicators was significantly reduced and the following variables included in the model for further analysis.

NSR: Russia's GDP (x_t^1) ; Number of vessels passing through the route (x_t^2); Average cost of passage through the route (x_t^3).

Southern Sea Route: GDP of the EU (x_t^1); Number of vessels passing through the route (x_t^2); Average cost of passage through the route (x_t^3).

Trans-Siberian Railway: Russia's GDP (x_t^1); Number of containers transported along the Trans-Siberian Railway (x_t^2); Average cost of transportation along the Trans-Siberian Railway (x_t^3).

Northwest Passage: Canada's GDP (x_t^1); Number of vessels passing through the passage (x_t^2); Average cost of passage through the route (x_t^3).

3.4. Time Series Analysis

Stationary analysis was carried out using the Dickey-Fuller test. This showed that all time series have a unit root. Therefore, the first differences form a stationary time series. Autocorrelation analysis of the endogenous and exogenous variables was performed in order to determine the lags for exogenous variables in the autoregressive model. Using the autocorrelation coefficients as a criterion, lags that are statistically significant are included in the model.

In the model for the NSR the endogenous variable was included with $t-1$ lag and two exogenous variables were included with $t-1$ and $t-2$ lag $\left(x_{t-1}^{1}, x_{t-2}^{1}\ and\ x_{t-1}^{2}, x_{t-2}^{2}\right)$.

In the model for the Southern Sea Route, the endogenous variable was included with $t-1$ and $t-2$ lag. Two exogenous variables were included with $t-1$ and $t-2$ lag $\left(x_{t-1}^{1}, x_{t-2}^{1}\ and\ x_{t-1}^{2}, x_{t-2}^{2}\right)$.

In the model for the Trans-Siberian Railway the endogenous variable was included with $t-1$ and $t-2$ lag. Two exogenous variables were included with $t-1$ and $t-2$ lag $\left(x_{t-1}^{1}, x_{t-2}^{1}\ and\ x_{t-1}^{2}, x_{t-2}^{2}\right)$.

In the model of the Northwest sea passage, the endogenous variable was included without a lag because of a small correlation coefficient between y_t and y_{t-i}. The exogenous variables were included with $t-2$ lag $\left(x_{t-2}^{1}\ and\ x_{t-2}^{2}\right)$.

The significance was checked using standard error and Q-Box-Pearson criteria.

Multicollinearity analysis revealed the following pattern. The correlation coefficients in the models for all corridors showed a close relationship between the volume of transported goods and GDP (0.89; 0.82; 0.69; 0.65 at $r = 0.9$).

The correlation coefficients between the volume of transported goods and the average cost of passage through the route do not show close relationship (r = 0.9). The correlation coefficients in the model for Southern Sea Route, Trans-Siberian Railway, NSR, and Northwest Passage are -0.0967, -0.0076, -0.215, and -0.88 respectively.

The coefficients of equation (1) were calculated for the stationary series. Based on the regression model (1), the equation with the remaining variables after the analysis is as follows:

$$y_t = a_0 + a_1 \cdot y_{t-1} + a_2 \cdot y_{t-2} + b_1\ x_{t-1}^{1} + b_2 \cdot x_{t-2}^{1} + c_1 x_{t-1}^{2} + c_2 x_{t-2}^{2} \tag{2}$$

The coefficients of the models for each corridor were determined using ordinary least squares. Student's t-test was used to assess the statistical significance of the coefficients.

Model of NSR:

$$y_t = 2.415 + 0.01 y_{t-1} - 0.05 x_{t-1}^{1} + 0.07 x_{t-2}^{1} + 0.007 x_{t-1}^{2} - 0.045 x_{t-2}^{2} - 0.13 x_{t-2}^{3} \tag{3}$$

The significance level of the coefficient of the variables x_{t-2}^{1}, x_{t-1}^{2}, is 0.696 and 0.864 respectively. Thus, these variables may be removed from the equation.

Model of Southern Sea Route:

$$y_t = 733.477 + 0.069 y_{t-1} + 0.016 y_{t-2} - 0.026 x_{t-1}^{1} - 0.025 x_{t-2}^{1} + 95.192 x_{t-1}^{2} + 13.198 x_{t-2}^{2} - 0.017 x_{t-2}^{3} \tag{4}$$

The significance level of the coefficient of the variable x_{t-2}^{2} is 0.794. Therefore, this variable may be removed from the equation.

Model of Trans-Siberian Railway:

$$y_t = 642457.687 + 80.743 y_{t-1} + 0.169 y_{t-2} - 0.214 x_{t-1}^{1} - 1.076 x_{t-2}^{1} + 259.0611 x_{t-1}^{2} + 45.211 x_{t-2}^{2} - 3.01 x_{t-1}^{3} \tag{5}$$

The significance level of the coefficient of the variable x_{t-2}^{2} is 0.873. Thus, this variable may be removed from the equation.

Model of Northwest Sea Passage:

$$y_t = 0.261 - 0.017 x_{t-2}^{1} - 0.143 x_{t-2}^{2} \tag{6}$$

The significance level of the coefficient of the variables is 0.99.

3.5. Conclusion

Climate change in the northern regions leads to ice melting and thus to a greater availability of routes in the Arctic Ocean in terms of available days of passage. This trend suggests that the number of vessels on the NSR and the Northwest sea passage will increase, as will, consequently, the volumes of transported goods. The NSR is the shortest path and the shortest sea transit corridor between Northern Europe and the Asia-Pacific region, which runs through the seas of the Arctic Ocean (the Barents-, Kara, Laptev-, East Siberian- and Chukchi Seas) and part of the Pacific Ocean (Bering Sea).

The NSR and the Northwest Sea Passage can become cost-effective alternatives due to the emergence of large-scale transportation, further development of the territory and a reduction in transportation costs.

The transportation corridor models allow the volume of traffic us to be estimated depending on various factors that characterise the different operating conditions of transport corridors, including natural, organisational, technological and economic factors. The results of the research outlined in this paper show that at present a lack of information on many indicators represents the biggest limitation to a full-scale operational modelling of the corridors. With the development of the Arctic zones proceeding apace, this suggests the necessity for a database of information to support analysis and forecasting.

Acknowledgements

This paper is based on research carried out with the financial support of the grant of the Russian Science Foundation (Project No. 14-38-00009). Peter the Great St. Petersburg Polytechnic University.

References

[1] Didenko N I et al 2017 Analysis of convergence-divergence in the development of economic processes in circumpolar countries *SGEM 2017 Political sciences law finance international relations* 537-545

[2] Giguère M A, Comtois C and Slack B 2017 Constraints on Canadian Arctic maritime connections case studies on transport policy **5 (2)** 355-366

[3] Aksenov Y et al 2017 On the future navigability of Arctic sea routes: high-resolution projections of the Arctic Ocean and sea ice *Marine policy* **75 (1)** 300-317

[4] Bardal K G 2017 Impacts of adverse weather on Arctic road transport *Journal of transport geography* **59 (2)** 49-58

[5] Didenko N I et al 2017 Deforestation and human activity: a global perspective *International multidisciplinary scientific geoconference surveying geology and mining ecology management SGEM 2017* **17 (53)** 183-190

[6] Didenko N I, Skripnuk D F and Mirolyubova O 2017 The effects of human behavior on fresh water resources *International multidisciplinary scientific geoconference surveying geology and mining ecology management SGEM 2017* **17 (53)** 901-910

[7] Mirolyubova O et al 2017 Analysis of world greenhouse gas emissions caused by agreeculture *International multidisciplinary scientific conference on social sciences and arts SGEM 2017* **IV** 57-64

[8] Rathi A K, Church R L and Solanki R S 1992 A macro level analysis of the airlift deployment problem *Computers & Operations Research* **19 (8)** 731-742

[9] Young O R 2016 Governing the Arctic Ocean *Marine Policy* **72 (10)** 271–277

[10] Gulas S et al 2017 Arctic Ocean oil and gas developments: Opportunities to improve governance and environmental pollution control *Marine Policy* **75 (1)** 53-61

[11] Bayerz M 2011 Legal status of the Northwest Passage and Arctic sovereignty Canada *Bulletin of Moscow University* 92-128

[12] Rodrigue J P 2009 The Northern East-West Freight Corridor (Eurasian Landbridge) *Hungarian Geographical Bulletin* **65** 3-15

[13] Bardeeva P 2011 Russ and Northern Sea Route in VI-XIII centuries: foreign policy and socio-economic aspects *Moscow state mining University* 9-13

[14] Dudarev G et al Advantage Northwest Russia. The new growth centre of Europe, Helsinki. Sitra. East-West Transportation Corridor (EWTC2)

[15] Lumiste R and Prause G 2013 Baltic states logistics and the East-West transport corridor Available from: www.ewtc2.eu.6 [Accessed 10th May 2017]

[16] Dodin D A 2005 Sustainable development of the Arctic: problems and prospects *St. Petersburg.: Science* 283

[17] Konyshev V and Sergunin A 2014 Is Russia a revisionist military power in the Arctic? *Defense & Security Analysis* **30** 323-335

[18] Mazur I I 2014 The Arctic – bifurcation point in the development of the global world, where the age of globalisation is moving ? p 306-317 (Volgograd: Teacher)

[19] Tennberg M 2017 Arctic environmental cooperation: a study in governmentality p 178 (University of Lapland)

[20] Didenko N, Skripnuk D and Mirolyubova O 2017 Modeling the changes in global temperature due to pollution. *International multidisciplinary scientific geoconference surveying geology and mining ecology management SGEM 2017* **17 (53)** 577-586

[21] Skripnuk D et al 2017 The formation of the regional tourist cluster in the Euro-Arctic Region *Int. Mult. Sc. Conf. on Social Sciences and Arts, SGEM 2017* **IV** 369-374

Modelling the Effect of Human Activity on Fresh Water Extraction from the Earth's Reserves

K N Kikkas[1], S V Kulik[12]

[1]Peter the Great St. Petersburg Polytechnic University, St. Petersburg, Russia
[2]St. Petersburg State University, Russia

xekikkas@gmail.com

Abstract. This paper reports on dependencies in changes to fresh water extraction patterns on the various influencing factors. The aim of the study is to develop a mathematical model to predict the influence of human activity on fresh water extraction from the Earth's reserves. The historical background of the process affects the current estimate for fresh water extraction. The current estimate of fresh water extraction is affected by various dynamic factors in human life as well as environmental processes. As a theoretical model for analysing the influence of human activities and their impact on the environment, the authors selected the autoregressive distributed lags model (ADL-model), in which the current values of the series depend both on past values of the series as well as current and past values of other time series. The authors described the methodology of empirical tests of the model. The paper presents the methods used to construct the model and approaches to determining model parameters. Statistical information for the model variables was collected from 1995 to 2016. Empirical testing of the model was carried out by the authors.

1. Introduction

Water is one of the simplest and most amazing substances in nature, existing in solid, liquid and gas states in nature – ice/snow, rivers/oceans and vapour, respectively. Water is present in all metabolic processes in living nature. Water is an inert solvent, i.e., it is not changed under the influence of the substances it dissolves.

About 71% of the Earth's surface is covered with water (oceans, seas, lakes, rivers, ice) and approximately 96.5% of the Earth's water is contained in the oceans. Around 1.7 % of global fresh water reserves is made up of groundwater, 1.7% of it forms glaciers and the ice caps of Antarctica and Greenland. A small part is in rivers, lakes and swamps, and 0.001% makes up the clouds. Thus, most of the Earth's water is salty and unsuitable for agriculture and drinking. Only 2.5% is fresh water, and 98.8% of this is in glaciers and groundwater. Less than 0.3 % of all fresh water is contained in rivers, lakes and the atmosphere, and even less (0.003 %) is in living organisms [1]. Globally, about two thirds of all precipitation returns to the atmosphere.

The most significant renewable water resources belong to the following countries: Brazil (8233 km^3); Russia (4508 km^3); USA (3069 km^3); Canada (2902 km^3); China (2840 km^3); Colombia (2132 km^3); Indonesia (2019 km^3); Peru (1913 km^3); India (1911 km^3); Venezuela (1320 km^3) [2]. Looking at water resources, the region of Latin America, which accounts for a third of the global water runoff is the richest, followed by Asia with a quarter of the world runoff. OECD countries possess 20%, while sub-

Saharan countries and the countries of the former Soviet Union have 10%. The most restricted in their water resources are the Middle East and North America (1% each).

Water stress is most intense in Sub-Saharan Africa; however, the overall water supply situation in the world is tense. Experts suggest that the volume of fresh water consumed by humankind will be equal to its reserves between 2035 and 2045. Both a threat of armed conflict caused by the lack of water supplies and the threat of social conflicts in separate countries may occur. Thus, a water shortage, which is turning into a global resource crisis, becomes literally a matter of health and of human life.

The aim of the study is to develop a mathematical model of the effect that human activity has on the fresh water extraction from the planet's reserves. This mathematical model should consist of an equation connecting the resulting (endogenous) variable and the influencing (exogenous) variables. The model will be used to analyse and predict the effect that human activity has on fresh water extraction. The model will detail how the endogenous variable will behave when the exogenous variables change.

The local objectives of the study are as follows:

- to describe the object of study — the extraction of fresh water under the influence of human activities;
- to choose the type of mathematical model;
- to outline the methodology of the empirical validation of the model;
- to validate the model using statistical data between 1990 and 2015.

2. Analysis of factors influencing fresh water extraction

Fresh water extraction is affected by various factors, which have been divided into three groups: human influence factors, environmental factors and the historical background factors of fresh water extraction processes.

The human influence factors: Human activity constantly promotes fresh water extraction, which takes various forms. Manufactures of coal, wood, pulp and paper industry, agriculture and utilities enterprises discharge harmful substances into water. Mercury, lead and their compounds are dangerous for the aquatic environment. Pollution affects not only surface water but also groundwater. In general, the groundwater condition is estimated as critical and has a dangerous tendency to further deteriorate. Underground water suffers from contamination caused by oilfields, mining, filtration fields, sludge ponds and dumps of metallurgical plants, storage of chemical wastes and fertilisers, landfills, livestock farms and inhabited settlements with no sewerage.

With environmental degradation and a growing population, water, once in abundance, virtually free and easy to use, is now becoming scarce and expensive.

About 25% of the water is used by industry processes and only 9% is used in the communal and household sector. According to statistics, most of the water resources are involved in agriculture (about 66% of all fresh water reserves). The main consumer of water is irrigation. To put this in perspective, growing 1 tonne of cotton requires 4-5 m^3 of fresh water and growing 1 tonne of rice needs 8 thousand m^3.

Household water consumption exceeds 20 cubic km per year globally. The level of development public water supply is defined by two parameters: the population provided with a centralised water supply and the value of specific water consumption. An important task is to reduce the consumption of tap water by industries. In Moscow, for example, industry accounts for 25% of tap water supplied to the capital. However, there is no need to use potable water for technical needs. If the network of industrial water supply expands, the cost of water overall will be reduced.

In total, industry uses about 90 cubic km of water per year. For example, smelting 1 tonne of steel requires 200-250 m^3 of water, 1 tonne of pulp needs 1300 m^3. Nevertheless, outdated petrochemical refineries consumed from 18 to 22 m^3 of water for refining of 1 tonne of oil, while modern plants with circulating water and air-cooling systems use about 0.12 m^3/year.

Environmental factors: The Earth is rich in overall fresh water resources; however, these are distributed unevenly. Meanwhile environmental degradation and climate change are also factors in the global fresh water crisis. Environmental degradation is reflected in the following processes:

climatological disasters; changes in CO_2 content; changes in greenhouse gas emissions from industry; changes in greenhouse gas emissions from agriculture; global temperature change; changes in stocks of fresh water and destruction (deforestation) of forest area.

Climatological disasters include: earthquakes; volcanic eruptions; avalanche; weather events; floods; landslides; fires; hurricanes; tsunamis.

The role of carbon dioxide in the biosphere is primarily the maintenance of photosynthesis, performed by plants. As a greenhouse gas, carbon dioxide in the air affects the heat exchange of the planet with the surrounding space, effectively blocking the radiation of heat on a number of frequencies, and thus participating in the formation and maintenance of the planetary climate. In connection with active human use of fossil energy as a fuel, there is a rapid increase in the concentration of this gas in the atmosphere. In addition, up to one third of total anthropogenic CO_2 emissions result from deforestation.

Greenhouse gases are gaseous constituents of the atmosphere, of natural or anthropogenic origin, that absorb and re-emit infrared radiation. Many types of industrial processes are connected with greenhouse gas emissions. The main sources of emissions are from industrial processes of chemical or physical processing of materials (e.g. blast furnaces in the steel industry; ammonia and other chemical products from fossil fuels used as chemical raw materials and cement production are the most important examples of industrial processes associated with the release of significant amounts of CO_2).

These processes produce a variety of greenhouse gases, including carbon dioxide (CO_2), methane (CH_4), nitrous oxide (N_2O), hydrofluorocarbons (HFCs) and perfluorocarbons (PFCs).

The largest source of greenhouse gas emissions in agriculture is enteric fermentation whereby methane is produced by animals during digestion and released into the atmosphere through regurgitation. Emissions from enteric fermentation increased by 11% between 2001 and 2011 [2]. The volume of emissions from the application of synthetic fertilisers accounted for 13% of all emissions in agriculture (the equivalent of 725 million tonnes of carbon dioxide) in 2011. Synthetic fertilisers are the fastest growing source of emissions in agriculture, having increased by 37% since 2001 [2].

One of the most obvious impacts of global temperature change is the shrinkage of forests. The "blame" is both with climate anomalies themselves (prolonged droughts, dry winters), and the outbreaks of pests and pathogens they cause. The shrinkage of forests causes great damage to forest ecosystems and produces conditions conducive to catastrophic wildfires.

Loss of fresh water due to reduced river flows are caused mainly by deforestation, cultivation of grasslands and drainage of floodplain wetlands. This leads, firstly, to increased surface runoff and increased volume of water flowing into the sea, and, secondly, to a decrease in groundwater levels that feed and support rivers. Loss of fresh water in many countries of the world reduces groundwater reserves. The environmental ecosystem is consequently affected, leading to a natural reduction of forest area.

In analysing factors affecting fresh water extraction processes, the history of the process is also relevant, i.e. the extraction of fresh water in the past affects the extraction of future fresh water. The mobility of water in the atmosphere and poor management of water resources can explain the influence of the previous period on fresh water extraction in a future period.

In the planet's atmosphere, water takes the form of small droplets, grouped into clouds and fog, as well as steam. In condensation, it is removed from the atmosphere in the form of precipitation (rain, snow, hail, dew). The condensation volume of the previous period affects the precipitation of the future period. Water, which undergoes the most global movement, after air, is constantly moving and travels over very long distances. When exposed to the sun's heat, water evaporates from plant surfaces, soil, rivers, ponds and seas. This forms water vapour, which is collected into clouds and transported by the wind before falling onto different continents in the form of snow or rain.

3. Literature review

The problem of water as an important element of human life causes scientific disputes between economists, lawyers, geographers and anthropologists [3]. The research examines the fresh water crisis as a new environmental crisis of the 21st century. Scientists emphasise [4,5] the need to consolidate the status of fresh water in international law and to conduct a comprehensive analysis of water law from the perspective of climate impacts [6].

Various authors analyse the use of fresh water for agriculture and problems arising in connection with this. The study [7] considers the introduction of institutional mechanisms, such as the introduction of quantitative water prices at the level of individual groups, replacing the existing pricing with a different approach in regard to the areas in question.

The study also considers how the fresh water scarcity affects the development of agriculture [8]. The paper presents the estimation of the benefits of irrigated agriculture as a result of access to different sources of water (water portfolio). The results of the study show that the use of water resources of different quality for agriculture, on the one hand, negatively affects the value of agricultural land but, on the other hand, reduces the sensitivity of agriculture to factors related to climate.

The research compares the use of fresh groundwater in Russia and in the United States [9]. The authors demonstrate the use of fresh groundwater for public water supply in separate regions and cities. They show that significant reductions in groundwater increase the danger of subsidence. Models of ground and surface water systems are presented in their works [10.11].

One of the main factors influencing the extraction of fresh water is human activity. The dynamic factors of human life are described in terms of economic growth and population growth [12-14]; from the perspective of ecology [15,16]; from the perspective of different influencing factors, including increasing demand for water [17]. Models exist that show the impact of human activities on greenhouse gas emissions [18]. The research [19] studies the impact of hydrological changes on biodiversity. Various anthropogenic activities are threatening the biodiversity of rivers and the water systems associated with them. Fresh water ecosystems are central to the global water cycle, its healthy functioning and the sustainability of other ecosystems. Inefficient use and extraction of fresh water are mainly studied from the position of possible threats to humans [20-22]. These threats, along with the growing demand for water resources can aggravate the problem of sustainable development.

4. Methodology

As a theoretical model for the analysis of the influence and impact of human activities on the environment, the authors selected the autoregressive distributed lags model (ADL-model), in which the current values of the series depend both on past values of the series and on current and past values of other time series.

In general, the ADL model of a multiplicative kind is shown below:

$$y_t^6 = A_0 \prod_{i=1}^{m} a_i y_{t-i}^6 \prod_{\substack{i=1,m;\\ k=1,n}} b_i x_{t-i}^k \prod_{\substack{s=1N;\\ i=1,m}} c_i y_{t-i}^s \qquad (1)$$

where A_0 is a free equation member; y_t^6 – is the endogenous variable "extraction of surface fresh water " in the year t; y_{t-i}^6 – is the exogenous variable - a factor in the prehistory of the process of fresh water extraction in the year $t-i$; m – the number of lags of the endogenous variable "extraction of surface fresh water", all the exogenous variables are included into the model with the same number of lags; x_{t-i}^k – k – type exogenous variable – the factor of human life in the period $t-i$; k – the number of exogenous variables; y_{t-i}^s – s - type exogenous variable – environmental factor in the period $t-i$; a_i, b_i, c_i - the parameters of the equation that need to be defined.

The methods of determining equation parameters include the following main stages: inspection of stationarity of the endogenous and exogenous variable time series using the Dickey-Fuller test; checking the multicollinearity of the exogenous variables; selection of the endogenous variable's lags, which have a strong correlation with the variable value in the last period and testing the significance of autocorrelation coefficients using the Ljung-Box Q-test; the choice of exogenous variables that have a strong correlation with the endogenous variable's value in the last period and testing the significance of correlation coefficients; representation of the ADL model equation in its multiplicative and additive form; finding the equation coefficients using the least squares method; model and determination coefficient accuracy validation using the F-Fisher criterion; validation of the regression equation coefficients using Student's t-criterion; executing the endogenous variable forecast for two years.

5. Date and analysis of the primary data

Working with the input data consisted of two steps: a) data collection for the selected variables; b) the primary processing of endogenous and exogenous variables that correspond to the analysed process and reflect the essence of the problem.

Data collection: Data was collected from 1995 to 2016. The data contains the values of endogenous and exogenous variables for the corresponding year in the countries of the world. The data was taken from the following sources: Euromonitor Passport Database, http://www.euromonitor.com/; World Bank Open Data, http://data.worldbank.org/.

6. Empirical testing of the model

The empirical testing of the model consisted of the following steps according to the equation parameter determination methodology.

Inspection of the time series of the endogenous variable and exogenous variables and their stationarity was performed using the Dickey-Fuller test. The Dickey-Fuller test is the solution for each variable of autoregressive equations like the following:

$$y_t = ay_{t-1} + \delta_t, x_t = ax_{t-1} + \delta_t, \quad (2)$$

where y_t, x_t — are time series, and δ_t — is a mistake. If $|a| < 1$, then the series is stationary.

For the endogenous variable "loss of fresh water stock" $a = 0.6645$ with $t_{cal.} = 6{,}4290 > t_{tabl} = 1{,}729$. Therefore, the time series of the endogenous variable "loss of fresh water" is stationary.

Checking the exogenous variables – factors of human life.

For x_{1t} $a = 0.9619$ with $t_{cal.} = 18.709 > t_{tabl} = 1{,}729$. For the exogenous variable x_{2t} $a = 0.9605$ with $t_{cal.} = 54.096 > t_{tabl} = 1.729$. For the exogenous variable x_{3t} $a = 0.9601$with $t_{cal.} = 65.6115 > t_{tabl} = 1.729$. For the exogenous variable x_{4t} $a = 0.4121$ with $t_{cal.} = 1.8952 > t_{tabl} = 1.729$. For the exogenous variable x_{5t} $a = 0.4095$ with $t_{cal.} = 1.8499 > t_{tabl} = 1.729$. All series of exogenous variables – factors of human life activity – are stationary.

Checking the exogenous variables – environmental factors. For y^1_{t-i} $a = 0.0507$ with $t_{cal.} = 2{,}2032 > t_{tabl} = 1{,}729$. For y^2_{t-i} $a = 0.9353$ with $t_{cal.} = 47{,}0291 > t_{tabl} = 1{,}729$. For y^3_{t-i} $a = 0.3559$ with $t_{cal.} = 1{,}8682 > t_{tabl} = 1{,}729$. For y^4_{t-i} $a = 0.8185$ with $t_{cal.} = 10.7094 > t_{tabl} = 1{,}729$. For y^5_{t-i} $a = 0.2090$ with $t_{cal.} = 2{,}4627 > t_{tabl} = 1{,}729$. For y^7_{t-i} $a = 0.7677$ with $t_{cal.} = 8{,}5097 > t_{tabl} = 1{,}729$. All series of exogenous variable - environmental factors are stationary.

Checking whether the exogenous variables are multicollinear: The pair correlation coefficients of indicators "y^2_t - CO_2 emissions per production unit", "y^4_t - Greenhouse gas emissions caused by agriculture", "y^1_t - Climate disaster" with other variables more than $|0.7|$, therefore, these factors will not be included in further analysis. Moreover, with the significance level α=0.05 and the number of degrees of freedom n-2=18, $t_{tabl} = 1{,}734$ [TB1]. For all the coefficients of pair correlation t – Student criterion $t_{cal} > t_{tabl}$ =1,734, i.e. the coefficients of pair correlation for the selected variables are significant.

The choice of the endogenous variable lags, which have a strong correlation with the variable value in the previous period and testing the significance of autocorrelation coefficients by means of Ljung-Box Q-test.

Analysis of correlation of the endogenous variable with the exogenous variable showed a strong correlation with y_t^1 (correlation coefficient 0.8661), y_t^2 (correlation coefficient 0.7184), y_t^3 (correlation coefficient 0.7619), y_t^4 (correlation coefficient 0.8278), y_t^5 (correlation coefficient 0.8990), y_t^7 (correlation coefficient 0.7190).

Variables $y_t^3, y_t^5, y_t^7, x_{1t}, x_{2t}, x_{3t}, x_{4t}, x_{5t}$ influence the endogenous variable y_t^6. The other coefficient values for the variables are less than 0.7, therefore they will not be included into further analysis. With the level of significance α=0.05, the number of freedom degrees n-2=16, t_{tabl}=1,746, and on all the cases when $t_{cal} > t_{tabl}$, i.e., the coefficients of pair correlation for the selected variables are significant.

The structural form of the ADL model of the multiplicative kind is:

$$y_t^6 = \alpha_0 \cdot y_{t-1}^{6^{\alpha_1}} \cdot y_t^{3^{\alpha_2}} \cdot y_t^{5^{\alpha_3}} \cdot y_t^{7^{\alpha_4}} \cdot x_{1t}^{\alpha_5} \cdot x_{2t}^{\alpha_6} \cdot x_{3t}^{\alpha_7} \cdot x_{4t}^{\alpha_8} \cdot x_{5t}^{\alpha_9} \quad (3)$$

Having made the left and right part of the equation logarithmic and introducing new variables, the structural form of the ADL model is written in the additive form:

$$A_1 = a_0 + a_1 A_{1t-1} + a_2 A_{3t} + a_3 A_{5t} + a_4 A_{7t} + a_5 B_{1t} + a_6 B_{2t} + a_7 B_{3t} + a_8 B_{4t} + a_9 B_{5t} \quad (4)$$

The model coefficients are found using the least squares method.

The significance of the regression equation and coefficients of the regression equation were verified by Fisher's F-criterion and Student's t-criterion. With α=0.05, F_{tabl}=30.144, F_{cal}=52,8449215 one can see that $F_{cal} > F_{tabl}$. Besides, the determination coefficient $R^2 = 0.932 > 0.7$, therefore, the regression equation is significant.

The equation with the obtained regression coefficients in its additive form:

$$A_1 = -3{,}09 + 0{,}24 A_{1_{t-1}} - 0{,}38 A_{3_t} - 0{,}185 A_{5_t} + 0{,}085 A_{7_t} + 1{,}997 B_{1_t} - 2{,}867 B_{2_t} + 1{,}163 B_{3_t} + 0{,}427 B_{4_t} - 0{,}321 B_{5_t}$$ [TB2] (5)

The equation in its multiplicative form:

$$y_{6_t} = e^{-3{,}09} \cdot y_{6_{t-1}}^{0.24} \cdot y_{3_t}^{-0.38} \cdot y_{5_t}^{-0.18} \cdot y_{7_t}^{0.085} \cdot x_{1_t}^{1.997} \cdot x_{2_t}^{-2.867} \cdot x_{3_t}^{1.163} \cdot x_{4_t}^{0.427} \cdot x_{5_t}^{-0.321} \quad (6)$$

7. Results and Discussion

We shall conduct analysis on the obtained parameters of the equation based on the following properties of the production function: the elasticity coefficient factor shows the percentage that the function will increase by if the factor increases by 1%. A positive elasticity coefficient means that the rise (fall) of a factor causes a corresponding increase (decrease) of the function.

Increase (decline) of the rate y_{t-1}^6 (loss of fresh water storage) by 1% in the previous period leads to an increase (decrease) of fresh water reserves by 0.24% in the current period. This process was termed a "background process". Similarly, the following factors influence the analysed endogenous indicator: y_t^7 growth (reduction of the forest area) by 1% leads to decreasing the fresh water supplies by about 0.085%; the growth of y_t^3 indicators (CO_2 emissions from industry), y_t^5 (the increase of global temperature), x_{2t} (increase in employment), x_{5t} (increase in imports) by 1% leads to decrease of fresh water by 0.38;0.18;2,86;0.32 percent, respectively.

The fresh water extraction is most of all influenced by factors of production activities: total world gross product (x_{1t}), the increase in the economically active population (x_{3t}). The growth of x_{1t} factor by 1% causes an increase of the function at 1.997%. The growth of x_{1t} factor by 1% causes an increase of the function by 1.163%.

The value of the constructed function, "the impact of human activities on the extraction of fresh water" lies in the fact that we can find out the maximum value of the function, which can be deduced in a given year with definite sizes and with a definite number of exogenous factors. The structure of the function "the impact of human activities on the fresh water extraction" remains constant over time. It can be adjusted to work both for very short periods of time, and for longer ones. The principle of adjustment is used in order to upgrade the function in regard to increasing accuracy.

8. Conclusions

A mathematical model for analysing the impact of the background of water reduction on the planet, the effects of human activities and the impact of external environment on the fresh water extraction on the planet is developed and presented. The endogenous variable of the model is associated with the exogenous variables, which evaluate the influencing processes. This paper presents the methods of constructing the model and shows how to find model parameters. Empirical testing of the model was carried out with statistical data from the period of 1995 – 2016. Further research is needed to improve the quality of modelling the impact of human activities on fresh water extraction.

Acknowledgements

This paper is based on research carried out with the financial support of the grant of the Russian Science Foundation (Project No. 14-38-00009). Peter the Great St. Petersburg Polytechnic University.

References

[1] Swain A 2008 Handbook of Water Resources in India: Development, Management, and Strategies *European Review of Agricultural Economics* **35 (4)** 589–591 (Oxford University Press)

[2] Demin A P 2014 Water resources and food program *Water Resources* **41 (3)** 232–241

[3] Vido S 2014 International Law for a Water-Scarce World *European Journal of International Law* **25 (3)** 935-941

[4] Chazournes L B 2013 Fresh Water in International Law p 288 (Oxford University Press, Oxford)

[5] Weiss E B 2013 International Law for a Water-Scarce World *Hague Academy of International Law Monographs* 344 (Martinus Nijhoff Publishers)

[6] Thielbörger P 2014 The Right(s) to Water. The Multi-Level Governance of a Unique Human Right chapter 2 190–219 (Heidelberg, Springer)

[7] Kajisa K, Dong B 2015 The Effects of Volumetric Pricing Policy on Farmers' Water Management Institutions and Their Water Use: The Case of Water User Organization in an Irrigation System in Hubei *World Bank Econ Rev* **31 (1)** 220-240 (China)

[8] Mukherjee M, Schwabe K 2015 Irrigated Agricultural Adaptation to Water and Climate Variability: The Economic Value of a Water Portfolio *American Journal of Agricultural Economics* **97 (3)** 809-832 (USA)

[9] Zektser I S, Karimova O A and Chetverikova A V 2015 Fresh groundwater resources and their use in emergencies *Water Resources* **42 (4)** 405–419

[10] Gemma M and Tsur Y 2007 The Stabilization Value of Groundwater and Conjunctive Water Management under Uncertainty *Applied Economic Perspectives and Policy* **29 (3)** 540-548

[11] Pedro-Monzonís M et al 2016 Water accounting for stressed river basins based on water resources management models *Science of The Total Environment* **565 (15)** p 181-190

[12] Grossman G M and Krueger A B 1995 Economic growth and the environment *Q J Econ* **110 (2)** 353–377

[13] Rudenko D Y, Pogodaeva T V and Didenko N I 2015 Poverty alleviation strategies in the russian arctic zone regions *Mediterranean Journal of Social Sciences* **6 (1)** 32 -39
[14] Mugagga F and Nabaasa B B 2016 The centrality of water resources to the realization of Sustainable Development Goals (SDG). A review of potentials and constraints on the African continent *International Soil and Water Conservation Research* **4 (3)** 215-223
[15] Romashkina G F, Didenko N I and Skripnuk D F 2017 Socioeconomic modernization of Russia and its Arctic regions *Studies on Russian Economic Development* **28 (1)** 22-30
[16] Allan J D et al 2005 Overfishing of Inland Waters *BioScience* **55 (12)** p 1041-1051
[17] Shao L and Chen G Q 2016 Embodied water accounting and renewability assessment for ecological wastewater treatment *Journal of Cleaner Production* **112 (5)** 4628-4635
[18] Didenko N I, Skripnuk D F and Mirolyubova O V 2017 Urbanization and Greenhouse Gas Emissions from Industry *IOP Conf. Series: Earth and Environmental Science* **72** 012014
[19] Dudgeon D 2000 Large-Scale Hydrological Changes in Tropical Asia: Prospects for Riverine Biodiversity: The construction of large dams will have an impact on the biodiversity of tropical Asian rivers and their associated wetlands *BioScience* **50 (9)** 793-806
[20] Matthews N 2016 People and Fresh Water Ecosystems: Pressures, Responses and Resilience. *Aquatic Procedia* **6** 99-105
[21] Didenko N I et al 2017 Analysis of convergence-divergence in the development of economic processes in circumpolar countries *SGEM 2017 Political Sciences, Law, Finance, International Relations* 537-545 (Vienna, Austria)
[22] Didenko N I et al 2017 Analysis of Modernization Processes in the Arctic Regions of Russia in the Global Context *SGEM 2017 Political Sciences, Law, Finance, International Relations* 557-564 (Vienna, Austria)

Evaluation of Russian Arctic Foreign Trade Activity

N A Konakhina

Peter the Great St. Petersburg Polytechnic University, St. Petersburg, 195251 Russia

konahina_na@spbstu.ru

Abstract. An analysis of foreign trade activity in the Russian Arctic framed by a discussion of the climatic, social, demographic, natural, etc. characteristics of the region is presented. From this point of view, the following limitations apply: the complexity of the Arctic area; the need to find technological solutions that take into account the climatic characteristics; an absence of the usual conditions for sustainable development; the limitation of ecological systems in the creation and use of technologies needed for the development of foreign trade activity. Theoretical approaches to the problem of foreign trade activity of territories are discussed alongside a proposed methodology for the analysis of foreign trade activity and an assessment of the foreign trade activity index of the Russian Arctic regions. The analysis is based on official data of Federal State Statistics Service of Russian Federation from 2000 to 2015. The results of the research showed that priorities for Russian Arctic foreign trade should be prioritised according to three groups, consisting of 1) regions with a high level of foreign trade activity; 2) regions with a low level of foreign trade activity; 3) regions with the least developed foreign trade activity. Each group requires a specific foreign trade development strategy corresponding to its characteristics.

1. Introduction

Foreign trade activity has a great significance for social-economic development of country. Foreign trade activity promotes the competitive advantages for countries and regions. The inefficiency of Russia's foreign trade structure is exacerbated in the Russian Arctic due to the preponderance of the raw materials sector and high dependence of the Russian economy for imported goods and services [1, 2, 3]. At the present time, the foreign trade activities of the Russian Federation are subordinated to the economic situation and at the mercy of world markets and global political decisions. The economic crisis in Russia has demonstrated a necessity for structural changes in foreign economic activity and the formation of a competitive economy based on knowledge and innovation [4, 5, 6]. The purpose of the research is therefore to assess the foreign trade activities of Russian Arctic in order to forecast and create an economic development strategy for the Russian Arctic based around knowledge and innovation.

1.1. Russian Arctic characteristics

Article I. The territory of Russian Arctic covers nine Federal Subject and includes: Murmansk Oblast; Nenets Autonomous Okrug; Chukotka Autonomous Okrug; Yamalo-Nenets Autonomous Okrug; the municipal district of Vorkuta (Komi Republic); the Allaikhovsk, Anabarsk (Dolgan-Evenk), Bulunsk, Nizhnekolymsk and Ust-Jansk uluses (districts) of the Republic of Sakha (Yakutia); the urban district of the city of Norilsk in Taymyr Dolgano-Nenets Autonomous Okrug (Krasnoyarsk Krai); the municipal formations of Arkhangelsk, the Mezensk municipal district, Novaya Zemlya, the town of Novodvinsk, Onezhsk municipal district, Primorsk municipal district, Severodvinsk

(Arkhangelsk region) [7]. For the purposes of our study, the following eight Arctic Russian Federal Subjects were selected: Arkhangelsk Oblast, Krasnoyarsk Krai, Republic of Sakha (Yakutia), Komi Republic, Yamalo-Nenets Autonomous Okrug, Chukotka Autonomous Okrug, Nenets Autonomous Okrug and Murmansk Oblast.

The analysis has shown that all of the Arctic territories have certain similar characteristics:

- Climatic features. For Arctic areas, it is common to have harsh climatic conditions and spatial gaps between the concentrations of natural resources [8, 9].
- Demographic characteristics. Only 5% of the Russian population lives in the northern latitudes. Northern territories are characterised by low density of population and multi-ethnic composition of the indigenous population.
- "Special northern subtype". The peripheral position of the Arctic regions was defined as the remoteness from the economic and cultural centres. One of the results of this spatial isolation is the high cost of transport and other associated production factors [10].
- Extensive exploitation of natural resources of Russian Arctic in Soviet period predetermined spatial organisation of the economy of the northern territories. The reclamation of Russian Arctic is characterised by continuous movement into new commodity areas. Arctic zone is represented by regions with largest deposits of natural resources: the Arctic regions of Russia account for 70% of oil, 90% of natural gas, 60% of copper, 100% of platinum group metals, as well as significant deposits of barite and apatite [11].
- Migration from Arctic regions during the period of transition from administrative-planned to a market economy. The transition to a market economy was accompanied by a decline of investment in ships and port facilities, disruption in the supply of necessary products and a reduction in maritime trade along the Northern Sea Route. Companies that were unable to reorient their business to foreign markets (forestry and fishing industry, agriculture, construction) faced a difficult financial situation. All of the above caused a significant decline in employment and an outflow of population.

1.2. Theoretical approaches to foreign trade activity

The foreign trade of different countries implies a part of foreign economic activity and includes imports and exports. Problems associated with the development of foreign trade activity in national economies have been addressed by various theoretical approaches [12, 13]. The mercantilists believed that the government should support a positive balance of trade through tariffs, quotas and other trade policy instruments. A positive balance of trade, also known as a trade surplus, occurs when exports exceed imports. According to mercantilism, the wealth of one country, which depended on the quantity of gold and silver within the country, could only grow at the expense of another country. By contrast, classical economic theory held the wealth of nations may depend on the necessary skills for producing goods and services and not on the quantity of accumulated precious metals. Hence, the main challenge is not the extraction of gold or silver but the production based on division of labour, cooperation and government non-interference in the economy. Based on the idea of absolute advantage, theory of foreign trade means that countries export those goods which they produce at lower costs (countries have absolute advantage in the production of the goods) and import those goods which are manufactured in other countries at lower costs (absolute advantages belong to their trading partners). Theory of absolute advantage has a number of assumptions, such as: a) labour is the one factor of production; b) only two goods and partners involved in international trade; c) transport costs are zero. In accordance with this approach, export is a positive factor for the national economy because it provides foreign sales for additional products that cannot be sold in the domestic market.

The theory of comparative advantage states that some countries produce goods and services at lower opportunity cost than other countries. According to the theory of comparative advantage, an opportunity cost is a benefit that could have been enjoyed but was relinquished in consequence of deciding to take a different production approach. The defining characteristic of comparative advantage theory is that international trade will be mutually beneficial for both countries even if one country has absolute advantage in goods or services production. The theory of comparative advantage relies on the following assumptions: a) labour is the only one factor of production; b) only two goods are involved in international trade; c) transport costs are zero.

In the mid 1920s, the factor proportions theory was developed. According to this approach, two countries, two goods and two factors of production (such as labour and capital) are considered as involved in international trade. The country is an exporter of goods when there is a relative abundance of the factors necessary for producing necessary goods. If a country has a relative lack of these factors, it needs to import those goods.

Further research within the framework of international trade questioned factor proportions theory. Thus, in 1953, the American scientist V. Leontiev demonstrated that labour-surplus countries export capital-intensive products but capital-surplus countries export labour-intensive products. The Leontiev paradox has been given a rational explanation according to which the author failed to take into consideration the fact that labour can be divided into skilled and unskilled.

The current stage of economic development of the world economy has given impetus to new theories of international trade such as the theory of the product life cycle, the theory of technological gap and the theory of preference similarity.

2. Methodology

In order to conduct the research, the author suggested the following method which consists of four stages (Fig. 1).

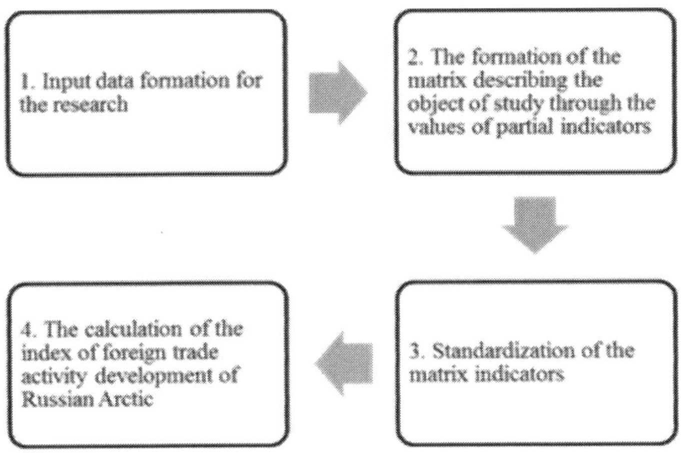

Figure 1. Method of evaluation the foreign trade activities of Russian Arctic

During the first stage, data for the object (Arctic regions of Russian Federation) is analysed.

During the second stage, the matrix (X_{ij}) is formed describing the object of study through the partial index (formula 1).

$$X = \|X_{ij}\| \; i = 1 \dots, n; j = 1 \dots, m \qquad (1)$$

where X_{ij}– value of j-th indicator of foreign trade activities for the i-th region;
n – number of analysed regions;
m – number of indicators of foreign trade activities for the regions.

The third stage is standardisation of the indicators included in the matrix (X'_{ij}). Since the indicators have different units, it is necessary to make a preliminary transformation which implies standardisation of indicators. The conversion is made according to formula 2:

$$X_{ij}^{/} = \frac{X_{ij} - \min_j(X_{ij})}{\max_j(X_{ij}) - \min_j(X_{ij})} \qquad (2)$$

where $X_{ij}^{/}$ – standardised value of j-th indicator of foreign trade activities for the i-th region;
X_{ij} – value of j-th indicator of foreign trade activities for i-th region;
$\max_j(X_{ij})$ – maximum of j-th indicator;
$\min_j(X_{ij})$ – minimum of j-th indicator.

In the fourth stage, the index of foreign trade activity development (I_{IT}) is calculated for each region (formula 3).

$$I_{IT} = \frac{\sum_j^m X_{ij}^{/}}{m} \qquad (3)$$

where I_{IT} – index of foreign trade activity development for the i-th region;
$X_{ij}^{/}$ – standardised value of j-th indicator of foreign trade activities for the i-th region;
m – number of indicators of foreign trade activities for the regions.

The index of foreign trade activity development is changed from 0 to 1 (I_{IT} = 1 is the maximum; and I_{IT} = 0 is the minimum).

3. Empirical specification

3.1. Data for the research

In order to carry out the research, statistical data were collected for the period from 2000 to 2015. The data were taken from the following sources: Federal State Statistics Service of Russian Federation.

The indicators show the state of foreign trade of Russian Arctic at the moment *t*:

1) Exports, US dollars (X_1);
2) Imports, US dollars (X_2);
3) Foreign trade quota (X_3);
4) Export quota (X_4);
5) Import quota (X_5);
6) Export-import coverage ratio (X_6).

3.2. Empirical specification

According to the proposed method, the author constructed two matrices for original and standardised indicators and calculated the index of foreign trade activity development.

Stage 1. The data were collected for the research.

Stage 2. A matrix was formed which describes the analysed object through the values of partial indicators (Fig. 2). A separate matrix was constructed for each period (*t*) from 2000 to 2015.

$$\begin{Vmatrix} X_{11} & X_{12} & X_{13} & X_{14} & X_{15} & X_{16} \\ X_{21} & X_{22} & X_{23} & X_{24} & X_{25} & X_{26} \\ X_{31} & X_{32} & X_{33} & X_{34} & X_{35} & X_{36} \\ X_{41} & X_{42} & X_{43} & X_{44} & X_{45} & X_{46} \\ X_{51} & X_{52} & X_{53} & X_{54} & X_{55} & X_{56} \\ X_{61} & X_{62} & X_{63} & X_{64} & X_{65} & X_{66} \\ X_{71} & X_{72} & X_{73} & X_{74} & X_{75} & X_{76} \\ X_{81} & X_{82} & X_{83} & X_{84} & X_{85} & X_{86} \end{Vmatrix}$$

Fig. 2. The matrix with the indicators by line (from X1 to X6) and regions by column (from X1 to X8) for period *t*

Stage 3. A standardised matrix was created for the period (*t*) from 2000 to 2015, describing the analysed object through the values of partial indicators (Fig. 3).

$$\begin{Vmatrix} X^{/}_{11} & X^{/}_{12} & X^{/}_{13} & X^{/}_{14} & X^{/}_{15} & X^{/}_{16} \\ X^{/}_{21} & X^{/}_{22} & X^{/}_{23} & X^{/}_{24} & X^{/}_{25} & X^{/}_{26} \\ X^{/}_{31} & X^{/}_{32} & X^{/}_{33} & X^{/}_{34} & X^{/}_{35} & X^{/}_{36} \\ X^{/}_{41} & X^{/}_{42} & X^{/}_{43} & X^{/}_{44} & X^{/}_{45} & X^{/}_{46} \\ X^{/}_{51} & X^{/}_{52} & X^{/}_{53} & X^{/}_{54} & X^{/}_{55} & X^{/}_{56} \\ X^{/}_{61} & X^{/}_{62} & X^{/}_{63} & X^{/}_{64} & X^{/}_{65} & X^{/}_{66} \\ X^{/}_{71} & X^{/}_{72} & X^{/}_{73} & X^{/}_{74} & X^{/}_{75} & X^{/}_{76} \\ X^{/}_{81} & X^{/}_{82} & X^{/}_{83} & X^{/}_{84} & X^{/}_{85} & X^{/}_{86} \end{Vmatrix}$$

Fig. 3. Standardised matrix with indicators by row (from X1 to X6) and regions by column (from X1 to X8) for period *t*

Stage 4. The author calculated the index of foreign trade activity development of Russian Arctic for the period from 2000 to 2015. Data are summarised in table 1.

Table 1. The index of foreign trade activity development of Russian Arctic for the period from 2000 to 2015

№	Regions of Russian Arctic	2000	2001	2002	2003	2004	2005	2006	2007
1	Murmansk Region	0.28	0.56	0.49	0.56	0.38	0.32	0.69	0.67
2	Krasnoyarsk Krai	0.70	0.62	0.61	0.71	0.52	0.37	0.72	0.73
3	Arkhangelsk Region	0.25	0.40	0.44	0.35	0.28	0.24	0.59	0.42
4	Komi Republic	0.37	0.51	0.59	0.58	0.27	0.29	0.35	0.34
5	Nenets Autonomous District	0.14	0.41	0.45	0.15	0.04	0.03	0.00	0.22
6	Yamalo-Nenets Autonomous District	0.35	0.57	0.63	0.65	0.17	0.15	0.22	0.14
7	Republic of Sakha (Yakutia)	0.39	0.51	0.44	0.44	0.44	0.42	0.66	0.50
8	Chukotka Autonomous District	0.00	0.00	0.05	0.12	0.78	0.78	0.42	0.36

№	Regions of Russian Arctic	**2008**	**2009**	**2010**	**2011**	**2012**	**2013**	**2014**	**2015**
1	Murmansk Region	0.58	0.62	0.38	0.55	0.59	0.66	0.64	0.59
2	Krasnoyarsk Krai	0.69	0.63	0.56	0.42	0.52	0.37	0.43	0.34
3	Arkhangelsk Region	0.52	0.82	0.62	0.42	0.39	0.31	0.36	0.39
4	Komi Republic	0.35	0.37	0.30	0.28	0.36	0.50	0.57	0.36
5	Nenets Autonomous District	0.26	0.17	0.57	0.61	0.53	0.48	0.33	0.33
6	Yamalo-Nenets Autonomous District	0.21	0.08	0.12	0.20	0.25	0.05	0.11	0.26
7	Republic of Sakha (Yakutia)	0.36	0.38	0.33	0.37	0.53	0.52	0.66	0.60
8	Chukotka Autonomous District	0.34	0.25	0.27	0.22	0.25	0.30	0.34	0.25

The results of calculation of foreign trade activity index are represented graphically in Figure. 4.

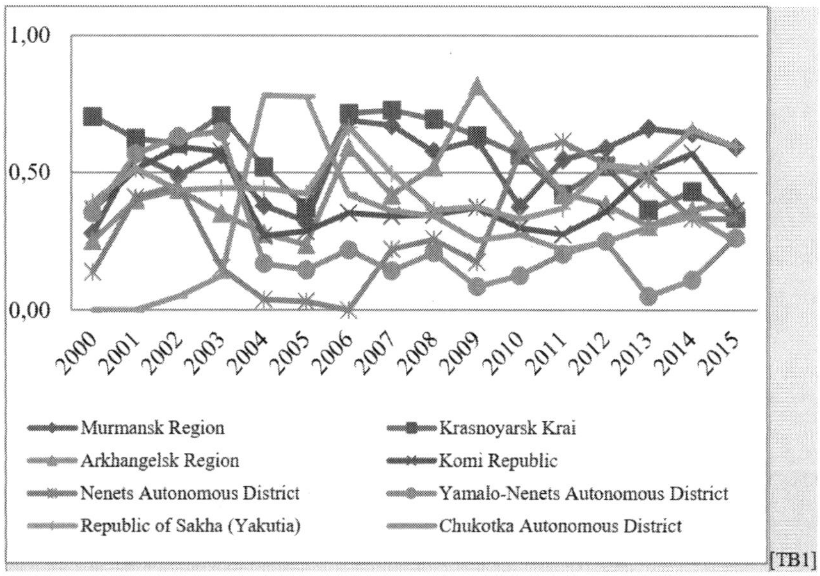

Figure 4. The index of foreign trade activity development of Russian Arctic for the period from 2000 to 2015

4. Conclusion

The analysis showed that there were significant variations in the level of development of foreign trade activities across the regions of the Russian Arctic. Thus, in 2015, the maximum index (0.6) was applied in the Republic of Sakha (Yakutia). The Murmansk Oblast demonstrates a slight lag from the maximum (0.59). Krasnoyarsk Krai, Arkhangelsk Oblast, Komi Republic and the Nenets Autonomous Okrug are at about the same level in terms of the development of foreign trade activities (0.34; 0.39; 0.36; 0.33).

The Yamalo-Nenets and Chukotka Autonomous Okrugs were the least developed in the sphere of foreign trade activities (0.26 and 0.25, respectively).

The results of research showed that Arctic regions may be divided into three groups as follows: 1) regions with a high level of foreign trade activity (Republic of Sakha (Yakutia) and Murmansk Oblast); 2) regions with low level of foreign trade activity (Krasnoyarsk Krai, Arkhangelsk Oblast, Komi Republic, Nenets Autonomous Okrug); 3) the least developed regions (Yamalo-Nenets and Chukotka Autonomous Okrugs). A specific strategy of foreign trade activity, which should be created for each group of Russian Arctic regions, will be the object of further research.

Acknowledgements
The paper is based on research carried out with the financial support of the grant of the Russian Science Foundation (Project No. 14-38-00009, The program-targeted management of the Russian Arctic zone development). Peter the Great St. Petersburg Polytechnic University.

References

[1] Rudenko D Y, Pogodaeva T V and Didenko N I 2015 Poverty alleviation strategies in the Russian arctic zone regions *Mediterranean Journal of Social Sciences* **6 (1)** 32–39
[2] Didenko N and Kunze K 2014 Relationship between Energy Policies and Quality of Life in Russia *WIT Transactions on Ecology and the Environment* **186** 3–1
[3] Romashkina G F, Didenko N I and Skripnuk D F 2017 Socioeconomic modernisation of Russia and its Arctic regions *Studies on Russian Economic Development* **28 (1)** 22–30
[4] Didenko N, Mirolyubova O, Skripnuk D and Kikkas K 2017 Model of the development of the Russian Arctic micro-territory *Int. Mult. Sc. Conf. on Social Sciences and Arts, SGEM 2017* **4** 517-524
[5] Skripnuk D and Ulitin V V 2016 Technical and economic substantiation of permafrost thermal stabilisation technology under global warming conditions *Materials Physics and Mechanics (Russia)* **26 (1)** p 85-88
[6] Pogodaeva T V, Zhaparova D V, Rudenko D Y and Skripnuk D F 2015 Innovations and Socio-Economic Development: Problems of the Natural Resources Intensive Use Regions *Mediterranean Journal of Social Sciences* **6 (1)** 129-135
[7] The presidential decree "About land territories of the Arctic zone of the Russian Federation. No. 296 dated 02.05.2014. Available from: http://www.kremlin.ru/acts/bank/38377 [Accessed 26th December 2015]
[8] Didenko N I, Skripnuk D F and Mirolyubova O V 2017 Urbanisation and Greenhouse Gas Emissions from Industry *IOP Conf. Series: Earth and Environmental Science* **72 012014**
[9] Didenko N, Rudenko D and Skripnuk D 2015 Environmental security issues in the Russian Arctic *Int. Mult. Sc. GeoConf. Surveying Geology and Mining Ecology Management, SGEM 2015* **3** 267-274
[10] Didenko N, Skripnuk D and Krasulina O 2016 Modelling the Spatial Development of the Russian Barents-Arctic Region *Int. Mult. Sc. Conf. on Social Sciences and Arts, SGEM 2016* **5** 471-478
[11] Rudenko D and Didenko N 2015 The System of Well-being Indicators in the Russian Arctic Regions *WELLSO 2015 - II Int. Scientific Sym. on Lifelong Wellbeing in the World, European Proceedings of Social and Behavioural Scienes (Russia)* **7** p 514-521
[12] Low K Y J and Ho E Y-C 2016 A Knowledge-based Theory of the Multinational Economic *Organisation Long Range Planning* **49 (6)** 641-647
[13] Iamsiraroj S 2016 The foreign direct investment–economic growth nexus *International Review of Economics & Finance* **42** 116-133

IOP Conf. Series: Earth and Environmental Science **180** (2018) 012019 doi:10.1088/1755-1315/180/1/012019

Problems of entrepreneurship development in the Russian Arctic Zone

O Y Krasulina

The Russian Presidential Academy of National Economy and Public Administration, Nizhegorodsky Institute of Management, Nizhny Novgorod, Russia

strash@mail.ru

Abstract. Due to its unique challenges, the Arctic region imposes the requirement of close cooperation in areas of private- and public-sector development. In the present work, the author aims to find the optimal ratio between state regulation and market mechanisms for the successful development of the energy sector Arctic region of the Russian Federation alongside improvements in the overall economic structure. Due to the complex impact on the system of interrelations between the Arctic unique set of regional resources, it makes sense to base the entire Arctic economy on the management of its natural resource potential. This will allow resources to be used more efficiently at the same time as satisfying resource-based needs and maintaining or even increasing the resource potential of the region. Since small and medium-sized enterprises (SMEs) react more quickly to various changes in the situation as well as being more resistant to sanctions and other external economic threats, focusing on their development is the most rational approach to the integrated socio-economic development of the Arctic region. A model of enterprises showing the dependence of production volumes on the factors that create them within municipal territories is proposed. This model is then used to as the basis for strategic tools for the development of the Arctic Zone of the Russian Federation.

1. Introduction

Under the challenging conditions of the Arctic Zone of the Russian Federation, it is necessary for government to support the implementation of investment projects by business interests operating in the region. Despite its rich natural resources, the cost of the products produced in the region remains high due to the complexity of the climatic conditions. For example, the costs of electricity generated by diesel power plants is quite high leading to annual budgetary losses in compensation for these tariffs. High port charges may add 10% to the cost of transported goods. Although close cooperation between federal and regional authorities is necessary for the implementation of investment projects in the Arctic region, the development of mineral resources continues to be hindered due to a lack of state support.

As the Russian experience of the 1990s has shown, the limitation of the functions of the state does not ensure the full implementation of market principles. [1] However, the Russian state remains the largest shareholder in the gas and oil industrial sector and problems concerning interaction between the state and private business remain to be resolved. When determining the correlation between state regulation and market mechanisms to optimally develop the energy sector and improve the structure of the overall economy, different stages of economic development, democratisation and changes in the political system need to be taken into account.

2. Analysis of the problems of Arctic exploration

The economy of the Arctic Zone varies greatly between constituent territories. Industrial, transport and port infrastructure objects – as well as labour resources – are placed extremely unevenly across the territorial and geographical context. While the western part of this zone is relatively developed and provided with suitable primary production infrastructure, the economies of other territories are subject to one-sided development with an underdeveloped infrastructure and shortage of material and labour resources. At the same time, it is necessary to consider that the innovative potential of the majority of the northern territories have suffer from a lack of scientific and educational resources. In these cases, special development mechanisms to support innovative development are necessary.

Under the existing Russian tax regime, the development of offshore fields in the Arctic Zone has a negative profitability. Even when taking temporary mineral extraction tax privileges into account, intervention measures on the Arctic shelf are ineffective [2]. The tax regime has a pronounced fiscal character and is rigidly focused on replenishing the state budget at any cost. In failing to promote the attractiveness of Russian and foreign investments in the oil industry or support the development of hard-to-recover reserves, extraction from low-profitable and low-yield fields and wells, this approach represents a departure from the resource management model.

Due to a failure to fully conform to modern standards, special measures for efficiency increases in the existing development of subsurface extraction are necessary. For example, the lack of a comprehensive overall geological-geophysical study of the domestic Arctic shelf for estimating actual hydrocarbon resources impairs the exploration and development of its full potential resulting in the Arctic territories being classed as highly or extremely risky from an investment perspective. [3] There are no clear, understandable and mandatory rules of subterranean resource exploitation. A basic adherence to commonly understood principles of oil and gas business include: a predictable and transparent licensing system, open tenders, tax breaks for companies in order to promote favourable conditions for attracting investment, ensuring the sustainability of the mineral resource base, a combination of regional and central interests and public control. [4]

The development of a competitive environment and antimonopoly regulation system applicable to the development of the Arctic shelf is an important strategic direction. While state regulation of geological exploration and production contributes little to the development of business activity, state-controlled monopolies have certain advantages. A focus on short-term financial profit does not create incentives to improve techniques and technology, and, consequently, to increase productivity and reduce costs.

Nowadays, the position of the Arctic Zone of the Russian Federation in the system of the international division of labour is defined mainly by the export of raw materials and hydrocarbons. [5] The structure and composition of exports indicate the limited competitiveness of an economy supported primarily by its diversity of natural resources. A quantitative approach to resource extraction results in the narrow specialisation of regional economies with a mono-profile orientation. Increasing reliance on raw materials significantly reduces the quality of growth. In the context of systemic global financial instability, investment opportunities for state and private business sectors have significantly decreased. The financing of many projects related to the integrated development of the Arctic Zone of the Russian Federation has been suspended and the terms of their implementation postponed indefinitely. At the current stage of development of the macro-region the problem of paramount importance is how to ensure economic adaptation including specific projects aimed at compensating for the negative consequences of changes in the global financial environment. In this regard, a positive role can be played by various forms of support of innovation-oriented industries. For example, it is possible to promote scientific research and experimental development by reducing the tax burden on R&D activities, forming a network of technology parks, business incubators and venture capital funds. [6] The leitmotiv of Arctic development should not consist only of large projects and new Urengoy and Norilsk, but also the development of an innovation-oriented environment together with the appropriate forms of business organisation and financing schemes. Here, a business environment based on the development of small and medium-sized businesses in various sectors of the economy is key. Moreover, the new gas fields which are located on the territory of the Yamal Peninsula, on the shelf of the Kara Sea and in the Pechora

Sea are fewer and more expensive to develop than those currently in development. According to such price volatility, which is typical for the gas and oil market, they are at the margins of profitability of large companies. Large corporations are only needed for operating super-efficient large deposits; the rest may be transferred to small companies.

3. Selection of strategic tools

In the course of developing the Russian Arctic, it is necessary in the first place to select the appropriate strategic tools. The development of the Arctic cannot be made to happen by employing methods of storm but instead demands realistic financing and calculation of long-term results. Therefore, it is critically important for the Arctic to acquire strategic planning that is not tied to the short-term situation of economic cycles. It is necessary to form scientific and technical groundwork, pay serious attention to exploration of reserves and create a reliable infrastructure to ensure economic activity and livelihoods.

Secondly, taking into account the key value for prospects of the region of production of natural resources, it is necessary to seek not only for extraction from this production of profit and tax deductions. The social value of extracted raw materials should be put at the forefront in the implementation of Arctic projects. The development of fields has to become a basis for the development of all regions with formation on this basis of a system of settlements, infrastructure facilities and local industry. Moreover, the development of the Arctic should be linked to the development of adjacent territories: the Urals, Siberia and the Far East.

Thirdly, the high costs for the development of economic activities in the Arctic require the concentration of resources. However, this must not imply an inevitable monopolisation of this activity by a small circle of corporations. On the contrary, state policy should aim to attract all interested Russian companies with the necessary competencies for Arctic projects. In particular, it is necessary to attract small and medium-sized businesses to this region. [7] The Arctic should become a region favourable for the use of innovative technologies and the cultivation of domestic technology companies.

Fourthly, for all projects implemented in the Arctic, there must be multiple sources of resilience from the point of view of environmental and human safety. Here it is not just that the self-healing nature of the violated ecological balance is extremely slow, resulting in possible costs to the companies involved. Nowadays, large-scale environmental accidents can also have extremely negative consequences from the point of view of external threats. [8] Russia's geopolitical rivals will certainly seek to use any such incidents to limit its sovereignty in the Arctic.

Fifthly, the scale and complexity of the problems associated with the development of the Arctic also requires a management system with appropriate complexity and diversity. On the one hand, an authoritative centre for the management of this region should be formed at the Federal level, having sufficient resources to address the large-scale tasks which are set before it. On the other hand, it is necessary to ensure a high degree of coordination between various agencies and regional authorities. In addition, territorial and sectoral management must be combined with project management.

There are high costs incurred in the Arctic Zone due to severe climatic conditions, remoteness, dependence on deliveries from other regions and increased salary expectations in connection with perceived northern privileges. However, in the specific severe economic conditions obtaining in the majority the Arctic regions, business revenues are lower than average across Russia of indicators. Table 1 presents revenue figures from the sale of goods (works, services – without VAT and excise duties) for SMEs on the territorial subjects of the Russian Federation (millions of rubles)[1]

Table 1 Proceeds from sales of goods (works, services – without VAT or excise duty)

[1]Here and further the Republic of Komi is not included in our analysis for the reason that only the city of Vorkuta belongs to the Arctic territories. Similarly, the Krasnoyarsk Territory is not included in our sample, because the Arctic is one Taimyr Municipal District. In all other cases of a partial assignment of a subject of the Russian Federation to the Arctic zone (the Arkhangelsk Region, the Republic of Sakha (Yakutia), there is a reference to the polar regions of at least three to five municipalities.

Region	per small and medium-sized enterprise operating in 2016	per small enterprise operating in 2016
Chukotka Autonomous district	52.9	34.9
The Yamalo-Nenets Autonomous District	41.2	28.6
Russian Federation	**37.2**	**30.6**
Murmansk region	33.2	27.4
Nenets Autonomous District	36.5	23.9
The Republic of Sakha (Yakutia)	26.4	21.4
Arkhangelsk region (excluding Nenets AD)	26.7	21.3

Source: The results of the continuous Federal statistical monitoring of the activities of small and medium-sized businesses in 2016 "Small business big country". M.: Rosstat. 2017.

4. Model showing the dependence of productivity on the factors that create it

Models of zero level are those applying to enterprises in the territory of municipal units. As a model of the enterprise, we will consider the production function. The production function is the economic-mathematical quantitative relationship between output quantities and production factors, such as resource costs, technology level.

One of the classic examples of the production function is Cobb-Douglas function, which shows the dependence of the volume of production (shipped products of the enterprise) on its factors of production-labour and capital costs (1).

$$Q = A \cdot L^{\gamma} K^{1-\gamma} \qquad (1)$$

where Q- the volume of shipped goods by enterprises,

A- technological coefficient (innovative development of the enterprise),

L- the number of employees,

K- capital,

γ- coefficient of elasticity by labour / capital.

The peculiarity of the industry – for example, YANAO – is that the largest share (about 90%) in the structure of shipped goods, completed works and services falls on the type of activity "Mining of natural resources". In the Autonomous region gas production is produced by 36 companies. The largest volume of the extracted gas is the share of subsidiaries of JSC Gazprom. Gas production is conducted by also small independent companies, such as LLC Novatek Tarkosaleneftegaz, LLC Yurkharovneftegaz, JSC LUKOIL-West Siberia.

Oil and gas condensate production was carried out by 38 enterprises.

Thus, the main specifics of the region consists of oil and gas production, which occupies overwhelming percent in the industry of YaNAO.

Each enterprise can be expressed through the production function:

$$Q_{=} f * A, L, K \qquad (2)$$

Q - the volume of the shipped production of the enterprise

A - technological coefficient (innovative development of the enterprise),

L - number of employees,

K - capital.

This production function is common. Of course, as for the Cobb-Douglas function, the functions for two sectors of the economy with the same technological coefficients will not be identical. For each enterprise there will be features of elasticity of dependence, which will be reflected through elasticity coefficients. However, since the bulk of the enterprises are engaged in development of minerals, this function will adequately reflect dependence of the production of the separate enterprises. [9] Thus, the model is a system of production functions, compiled for each enterprise.

Corporations working in the Arctic region cannot solve the economic and social problems of the region in a complex way [10-14]. To develop small and medium-sized businesses is the most rational for the comprehensive socio-economic development of the Arctic region. Such enterprises react more quickly to various changes in the market situation, and many of them are more resistant to crisis and sanctions phenomena in the economy.

5. Conclusion

In the Arctic region, the state has to act as the initiator and organiser of diverse platforms of active communication between businessmen of the Arctic autonomous districts on a regular basis. The severe climatic and economic working conditions in the Arctic Zone necessitate a revaluation of the existing tax law. The question is "In what forms and using which specific instruments should this manoeuvre be realised?" Most respondents suggest a decrease in insurance premiums and reducing income tax of organisations operating in the Arctic to 10% as well as to reduce property tax. It is proposed that more active use of the compensation tool from the Federal, regional and local budgets be made for payments to social funds for employees of innovative small and medium-sized enterprises of the Arctic. In addition, it is proposed to make active use of special tax regimes for entrepreneurs of the Arctic and to simplify the transition to them.

The state decision to attract talented young people to the Russian Arctic is still lagging behind. Small business can solve this problem more quickly. Special measures for encouraging youth business in the Arctic Zone are necessary. The Federal Corporation for enterprise development could provide a special package of incentive measures in this direction. Through business there can be primary involvement of talented and vigorous Russian youth to the Arctic which will be able to strengthen further the corporate and budgetary structures. However, the first step in solving this state problem is to provide more support for local small businesses.

Acknowledgements

This paper is based on research carried out with the financial support of the grant of the Russian Science Foundation (Project No. 14-38-00009). Peter the Great St. Petersburg Polytechnic University.

References

[1] Alekseev G V et al 2016 Russia's Arctic Space in the XXI century: Development Factors, organisation of management chapter 7 153–191

[2] Didenko N I, Skripnuk D F and Mirolyubova O V 2017 Big data and the global economy *10th Int. Conf. Management of Large-Scale System Development* 8109611

[3] Didenko N I and Komahina N 2017 Features of the Arctic Zone of the Russian Federation from positions of innovative development. *In the coll.: Comprehensive development of the Arctic Coll. of scientific works Of the international Symposium* 15-26

[4] Didenko N I and Kunze K 2014 Relationship between Energy Policies and Quality of Life in Russia. *WIT Transactions on Ecology and the Environment* **186** p 3-11

[5] Didenko N I, Skripnuk D F and Krasulina O Y 2016 Delling the spatial development of the Russian Barentsarctic region *Politicalsciences, law, finance, economics and tourism conference proceedings* **24-30** 471-478

[6] Didenko N I, Rudenko D and Skripnuk D F 2015 Environmental security issues in the Russian Arctic *Int. Mult. Sc. GeoConf. Surveying Geology and Mining Ecology Management, SGEM* **3 (5)** 267-274

[7] Romashkina G F, Didenko N I and Skripnuk D F 2017 Socioeconomic modernisation of Russia and its Arctic regions *Studies on Russian Economic Development* **28 (1)** 30–40

[8] Popkova A., Kostko N and Skripnuk D F 2017 The quality of social space mapping: The case of tyumen, Russia *Int. Mult. Sc. GeoConf. Surveying Geology and Mining Ecology Management, SGEM* **17(23)** 753-760

[9] Didenko N I, Skripnuk D F and Mirolyubova O V 2017 Urbanisation and Greenhouse Gas

Emissions from Industry *IOP Conf. Series: Earth and Environmental Science* **72** 012014
[10] Afenyo M et al 2017 A probabilistic ecological risk model for Arctic marine oil spills *Journal of Environmental Chemical Engineering* **5 (2)** 1494-1503
[11] Nilsen T 2017 Innovation from the inside out: Contrasting fossil and renewable energy pathways at Statoil *Energy Research & Social Science* **28** 50-57
[12] Asheim B, Grillitsch M and Trippl M 2017 Chapter 4: Smart Specialisation as an Innovation-Driven Strategy for Economic Diversification: Examples From Scandinavian Regions *Advances in the Theory and Practice of Smart Specialisation* book chapter 73-97 (Academic Press)
[13] Mirolyubova O et al 2017 Analysis of world greenhouse gas emissions caused by agreeculture *Int. Mult. Sc. Conf. on Social Sciences and Arts SGEM 2017* **IV** 57-64
[14] Skripnuk D et al 2017 The formation of the regional tourist cluster in the Euro-Arctic Region *Int. Mult. Sc. Conf. on Social Sciences and Arts, SGEM 2017* **IV** 369-374

IOP Conf. Series: Earth and Environmental Science **180** (2018) 012020 doi:10.1088/1755-1315/180/1/012020

Analysis of advanced nuclear technologies applicable in the Russian Arctic

Viktor Merkulov

Peter the Great St. Petersburg Polytechnic University, St Petersburg, Russia

merkulov.v@gmail.com

Abstract. The main areas and developmental prospects of nuclear technologies used in the Arctic Zone of the Russian Federation are considered in the article. The main directions of the key Russian nuclear research centres and development are discussed. An analysis of advanced nuclear research and development aimed at the development of the Arctic is carried out. The following advanced technologies applicable in the Russian Arctic were described in the article: low-power nuclear power plants; an autonomous automated nuclear power plant for the power supply of subsea and sub-ice drilling and oil and gas production complexes on the shelf of the Arctic seas; a transportable nuclear power plant with a VBER-300 reactor facility based on a stationary ice-resistant platform for power supply of the gas production and processing centres in the Arctic; a system of autonomous power supply of the remote control centre of the prospective hydrocarbon deposit of the Arctic shelf on the basis of an autonomous nuclear power unit; nuclear power modules for power supply of subsea and sub-ice facilities of the Arctic offshore fields; a nuclear subsea gas pumping station; the atomic icebreaker project 10510 – "Leader". Conclusions about priority areas of development for nuclear technologies in the Russian Arctic are formulated.

1. Introduction

In the present context of the intensive development of the Arctic Zone of the Russian Federation, the problem of ensuring fail-safe power supply becomes especially urgent. In a number of energy-deficit areas, electricity and heat supply problems can be solved by using currently available alternative energy sources. However, the severe natural and climatic conditions characteristic of the northern regions of Russia make it difficult to implement electricity generation solutions based on the transformation of solar, wind, wave or tidal energy. Conversely, the deployment of local generating facilities based on nuclear reactor facilities presents itself as a reliable, safe and cost-effective solution to the problem of maintaining electricity supply. At present, there are two nuclear power plants operating in the Russian Arctic – the Kola (1760 MW) and Bilibino (48 MW) facilities.

The year-round navigation through the Northern sea route by means of a fleet of nuclear icebreakers is a very important aspect of development of economic activity in the Russian Arctic. The development of new nuclear icebreaker projects will make it possible for ships having a larger deadweight and hull width than is currently possible to navigate the Arctic ice seas, as well as to increase the speed of vessels currently navigating severe ice situations.

2. Research and development centres of nuclear technologies applicable in the Russian Arctic

Among technological leaders in the field of nuclear research and development technologies for the Russian Federation are the following companies: JSC "Afrikantov OKBM", JSC "NIKIET", National Research Centre "Kurchatov Institute", JSC "Atomenergo".

2.1. JSC "Afrikantov OKBM"

JSC "Afrikantov OKBM", a large scientific and production centre of the "Rosatom" state corporation, is capable of full-cycle development: the manufacture and delivery of reliable, safe and efficient reactor plants of military and commercial applications, equipment and systems for reactor units and nuclear power stations, as well as their maintenance. Thus, Afrikantov OKBM is capable of carrying out a complete range of works to produce reactor units of various types and equipment for nuclear power plants, including the development of engineering documentation and associated calculations, research and development activities, manufacturing and testing of pilot items with serial improvement of production technology, supervision of the manufacture and assembly of standard equipment, commissioning and starting, maintenance and decommissioning.

For the market sector of regional electric power industry and the thermal power industry, Afrikantov OKBM has developed designs of reactor plants of VBER type (water-cooled modular power reactor) with a power rating from 100 to 600 MW(e). VBER reactor units incorporate an optimal combination of the latest achievements in marine reactor technologies together with proven solutions for the core and fuel cycle VVER-1000 and meet all basic safety, reliability and efficiency requirements for the prospective next generation of nuclear stations.

The list of advanced Arctic research and development of projects carried out by Afrikantov OKBM includes: nuclear power unit for underwater uninhabited nuclear power modules for underwater and under-ice supply of electricity to drilling and mining facilities; nuclear power source with electric power up to 6 MW in types of floating and block-transportable power units; transportable nuclear power plant with a VBER-300 reactor facility based on a stationary ice-resistant platform for power supply to Arctic gas production and gas processing centres; development of modern technical and economic requirements of nuclear ship power units based on empirical-numerical supercomputer technology [1].

2.2. JSC "NIKIET"

JSC "NIKIET", representing a multidisciplinary institute engaged in a number of major nuclear power engineering areas, is one of Russia's largest nuclear engineering and technology centres. The main areas of activities of the organisation comprise: the development of power reactors for nuclear power plants; the development of advanced power reactors for the nuclear power of the future, new nuclear fuel cycles, including fast reactors with a lead coolant with a closed-loop fuel cycle; the development, manufacture and testing of nuclear research facilities; starting, operation, decommissioning and utilisation of nuclear power plants, military and civil reactors and their ground-based prototype stands; operation of research nuclear reactors and facilities; research works in the areas of reactor physics, thermal physics, hydrodynamics, reliability, strength, materials science and corrosion; research and development, calculation and experimental, engineering and other works for ensuring the safety of nuclear power plants and research reactors, etc.

The list of NIKIET's advanced Arctic research and development projects includes: a power unit for a low-power nuclear power plant with a Unitarm-30 reactor unit; a power unit for a low-power nuclear power plant with a NIKA-330 reactor facility; a VRK type boiler reactor power line and designs of power units for low-power nuclear power plants based thereon; power unit of a low-power nuclear power plant with an installed SHELF reactor [2].

2.3. NRC "Kurchatov Institute"

The National Research Centre "Kurchatov Institute" is one of the leading research centres in the world and the largest interdisciplinary laboratory in Russia. A substantial number of Russian nuclear physics facilities have been consolidated in the Kurchatov Institute.

The main areas of activities of the organisation comprise: the organisation of research, development and engineering works in the fields of nuclear power, controlled thermonuclear fusion and plasma processes; participation in the engineering, construction, deployment, operation and decommissioning of nuclear and thermonuclear facilities and radioactive waste storage facilities; participation in the engineering and manufacture of equipment for nuclear and thermonuclear facilities, radiation sources, storage facilities for nuclear materials and radioactive substances.

The list of advanced Arctic research and development projects undertaken by the Kurchatov Institute includes: simulation model of power supply of the underwater complex of development of the gas condensate field of the Arctic shelf; an autonomous power supply system for a remote control centre for the development of a hydrocarbon field on the Arctic shelf on the basis of an autonomous nuclear power unit; atomic energy modules for power supply of subsea and sub-ice facilities for the Arctic offshore fields; scientific and technical problems of integrated environmental monitoring and forecasting of the state of the environment and objects in relation to promising deserted oil and gas production technologies in the Arctic and on the continental shelf [3].

2.4. JSC "Atomenergo"

JSC "Atomenergo" is a leading Russian engineering company in the field of civil nuclear power shipbuilding. The main areas of activities of the organisation include: creation of floating nuclear power plants; creation of means for handling spent nuclear fuel; engineering for nuclear technology service vessels; development of engineering documentation and supply of equipment for water treatment of nuclear power plants.

The list of advanced Arctic research and development projects by Atomenergo includes: the project of a Floating Nuclear Power Plant (FNPP) based on the ABV type reactor installation designed to provide coastal inaccessible areas with electric and thermal energy; a floating low-power nuclear power plant with KLT-40S reactor installations (Akademik Lomonosov FNPP); the transportable nuclear power plant project with a VBER-300 reactor facility based on a stationary ice-resistant platform for power supply to Arctic gas production and processing centres; the project of FNPP on the basis of power unit with RU KLT-40S for the city of Vilyuchinsk in the Kamchatka Region [4].

3. Analysis of advanced nuclear technologies for the development of the Russian Arctic

3.1. Low-power nuclear power plants (LPNPP)

In the medium-term, the role of low-power nuclear power units is the replacement of existing nuclear power generation capacities, also based on traditional types of fossil fuels, as well as the construction of local electricity generation centres for newly-created production and processing facilities in remote Arctic areas.

To date, NIKIET has developed a line of modern power units in the range of 0.2-100 MW of electric power. Among them are the following projects, which are currently at different stages of development: transportable low-power plant with a water-cooled Vityaz reactor with an electric capacity of up to 1 MW; transportable low-power plant on the basis of an innovative gas-cooled reactor of electric power up to 3.2 MW; UNITERM reactor facility for a stationary low-power plant [8]; a power unit with a SHELF reactor facility engineered as a power capsule with built-in reactor and turbo generator sets [5]; installation of integrated NIKA reactor for an LPNPP with an electric power of 100 MW. A pressure compensator placed in the reactor vessel at the station having a NIKA reactor minimises the risk of a LOCA type accident [8]; a line of hull boiling devices developed by NIKIET for stations with electric capacity of 45, 100 and 300 MW in single-block design.

In 2016, the feasibility of creating a transportable LPNPP on the basis of water and gas cooled reactor facilities for the purposes of locally-based power supply in remote locations was studied. The transportation is carried out in the form of transportable modules on a wheel-based platform. Engineering documentation of the water-cooled nuclear unit Vityaz was developed in 2016.

The concept of the transportable LPNPP with ATGOR gas-cooled reactor was developed using the results of the work on the high-temperature gas-cooled reactor project. The electrical power generated (0.2-8 MW) depends on the delivery set. The energy conversion system based on mass-produced gas turbine engines allows the use of traditional generation on organic fuel during maintenance stop of ATGOR.

The SHELF unified nuclear unit (NU) is developed on the basis of experience in the design, engineering and operation of existing facilities. Here, the type of power unit may either be ground-based or underwater, depending on the location. It is supplied in the form of an energy capsule with NU or with a compactly positioned NU and Atomic turbo generator located in a single shell. At present, the technical design and preliminary safety report materials (covering the stage of reactor installation) are being developed.

NIKIET is currently working on the formation of cooperation between manufacturers with the aim of producing of the main sample of LPNPP. Agreements on cooperation and interaction on LPNPP have been signed with the Kurchatov Institute [2].

3.2. Autonomous automated nuclear power plant for the power supply of subsea and sub-ice drilling and oil and gas production complexes on the shelf of the Arctic seas.

For the supply of subsea and sub-ice drilling and production facilities, OKBM Afrikantov has developed projects involving automated autonomous nuclear power plants with water-cooled pressure reactors (WCPR) with the use of compact sealed high-power turbine generators and nuclear power units with high-temperature gas reactors (HTGR) and a closed gas turbine cycle with a useful capacity from 8 to 25 MW of electric power. This type of nuclear power plant facility is able to operate for a long time without a human management or direct maintenance presence.

The chief technical characteristics of the nuclear power plant are as follows: electric power – 8 MW; depth of operation— 100-400 below sea level; total service life (without recharging fuel) – 30 years. Dimensions of the energy compartment: a. ∅ 11×13 m (for nuclear power plant with WCPR); b. ∅ 11×10 m (for nuclear power plants with HTGR). A block of modules with a total capacity of up to 250 MW or more can be assembled from single modules with a capacity of 8-25 MW of electric power. Reliable WCPRs developed by Afrikantov OKBM are used as nuclear power plants for the Russian sea vessels. Technical WCPR solutions from currently-operating and under-construction transportable nuclear units are proposed for use in drilling and subsea production complexes.

The proposed installations are engineered taking into account the use of a unified element base for equipment (active zones, steam generating elements, body equipment, pumps, CPS actuators, fittings, heat exchange equipment, etc.). Completed R&D and long-term operational experience of installations of this type at various transportable objects of energy ensure the reliability and resource characteristics, as well as the safety of the installation as a whole. In the proposed nuclear power plants, principles of self-management and self-protection, passive safety and autonomous responses in the case of accidents are developed to the maximum possible extent.

Nuclear power plants having a high-temperature gas reactor (HTGR) comprise an innovative technology, the main technical solutions of which are demonstrated in the work carried out by Afrikantov OKBM within the framework of certain programmes of the Rosatom state corporation. It should be noted that one distinguishing feature of nuclear power plants with HTGR is their high efficiency in comparison with water technologies, which makes it possible to increase the capacity of the unit within the limits of the specified overall characteristics of nuclear power plants. The development of gas technologies in the transport reactor building requires an expansion of the R&D complex including the creation of a domestic ground nuclear test facility for testing high-temperature gas coolant technology. According to preliminary estimates, a pilot plant with HTGR can be created within 8-10 years. Taking into account the innovations used in nuclear power plants with HTGR, this direction of NPP use in the objects of offshore zone development has priority in the long term [8].

In addition to the submarine modules intended for power supply of drilling and production complexes, high-temperature gas coolant technologies can be used for other underwater objects serving the shelf zone, for example, robotic autonomous units (observation and repair of drilling equipment, etc.), as well as low-power, deep-sea nuclear power plants.

3.3. Transportable nuclear power plant with a VBER-300 reactor facility based on a stationary ice-resistant platform for power supply of gas production and processing centres in the Arctic

The production of liquefied natural gas (LNG) requires significant energy supply. Here, electricity is utilised to satisfy technological requirements for the separation of produced gas into fractions and bringing methane – the main component of natural gas – into a liquefied state at atmospheric pressure and a temperature of minus 163 °C, in which state it is stored and transported by gas carriers. Traditional sources of energy in the LNG cycle include a gas turbine-generator, in which part of the produced well production is consumed.

Energy supply of LNG production capacity of 10 million tons requires the burning of approximately 1 billion m^3 of gas per year. Corresponding to this volume, the lost profit from the sale of such a volume of gas in the world market will be up to $ 180 million per year at a market price for natural gas of $ 200 per 1000 cubic meters, or $ 5.4 billion over 30 years. Concerning this issue, there is a question as to what extent it is reasonable to replace gas for internal technological needs of LNG with an atomic energy source. This circumstance indicates the need for a search for alternative sources of energy supply for satisfying the needs of offshore gas fields and for LNG production.

As such an alternative energy source, transportable nuclear power units are developed on the basis of a stationary ice-resistant platform with a VBER-300 reactor facility. Engineering of this type of nuclear power plant (NPP) (transportable, based on a stationary ice-resistant platform) is caused by a reduction in the cost of the facility in comparison with other options for nuclear power generation and takes into account the current trends in the organisation and location of LNG production. The electrical capacity of the NPP (300 MW) corresponds to the energy consumption of the LNG plant in the amount of approximately 5-5.5 million tonnes of LNG per year. As a rule, for large deposits, LNG plants are planned to be constructed in lines with a total capacity of 10-16.5 million tons. As a power source at NPPs, a compact reactor unit VBER-300, developed by Afrikantov OKBM based on shipboard nuclear power engineering technologies, is used [6].

Engineering the NPP to be based on a stationary ice-resistant platform has a number of advantages in comparison with execution in the form of floating nuclear power plants or a ground power unit: the use of floating NPPs limits the geographic location and requires the construction of protective hydraulic structures; the construction of a 300 MW nuclear power plant on the shore, even with the use of floating technology, is not optimal due to the large amount of construction and installation work performed directly on the site of the NPP, which, due to the inaccessibility of the construction site (the Arctic coast), leads to a significant rise in cost of project. The main dimensions of the stationary ice-resistant platform are: length – 132m; width – 132m; height – 24m; mass about 80,000 tons.

In case of the high market prices of gas and increase in gas extraction costs in extreme natural and climatic conditions at distant Arctic areas the use of the proposed transportable nuclear power plant for powering LNG facilities may be more beneficial than burning of expensive hydrocarbon fuel.

3.4. System of autonomous power supply of remote-control centre of prospective hydrocarbon deposit of the Arctic shelf on the basis of autonomous nuclear power unit

The remote-control centre of prospective oil and gas field is a highly effective system for monitoring and managing the development of the field. It takes into account the structural features of a field that provides a significant increase in the level of oil and gas output.

Such a remote-control centre presupposes the availability of all necessary equipment for remote monitoring and management of facilities for the development of subsea ice production. This approach is formed on the basis of advanced solutions used in contemporary Russian and foreign projects for transitioning to autonomous technologies in order to minimise the number of personnel employed directly at the field site with the use of highly automated and robotic technological systems and subsea and sub-ice production devices [7].

The distances to which subsea control cables along the shelf bottom can be laid today reach values of more than 200 km. The high-speed transmission of monitoring and control information using communication and power transmission technologies allows the power supply for offshore field development and production facilities to be carried out at high speed. This concept of autonomous power supply to subsea production facilities is possible due to highly automated underwater power systems that can be in close proximity to consumers for the implementation of their reliable and uninterrupted power supply. The use of an autonomous nuclear power unit with a capacity of 250 KW is proposed for providing power to the remote-control centre of the prospective North-Wrangel deposit [3].

3.5. Nuclear power units for power supply of subsea facilities of the Arctic offshore gas fields

The Kurchatov Institute has carried out the development of a nuclear power plant for direct energy conversion for a submersible power module within the framework of cooperation with JSC "SPMBM" Malachite". SPMBM Malachite proposed the concept of using a nuclear power plant for direct energy conversion (NPP DEC) as part of a submersible energy module (SEM) designed for servicing the power supply requirements of a complex of subsea and sub-ice facilities for oil and gas fields on the Arctic shelf. The design of the SEM with NPP DEC is proposed to be presented in the form of a permeable frame structure installed on the bottom of the platform base with ten "sockets" for the placement of the NPP PES. A robust, dry electro-technical compartment with electrical panels, transformers, control panel is installed on the platform base. The platform is located on supports above the ground bottom to prevent the bottom hydrants from moving out. The electrical compartment is equipped with unified international standard coamings to ensure the docking of habitable underwater vehicles. The premises and equipment of the electrical compartment on the SEM do not require long-stay personnel.

This approach is formed on the basis of promising future solutions used in modern Russian and foreign projects for the transition to autonomous technologies in the development of the field in order to minimise the number of personnel working directly at the field site with the use of highly automated and robotic technological systems and subsea mining devices. The SEM with NPP DEC can be installed at a depth of up to 400 m and provide power supply for technological systems and underwater ice facilities with a capacity of up to 10 MW.

For safe and reliable power supply, a SEM with a NPP DEC includes in its project technical solutions and measures based on the principles of natural safety (natural circulation of coolants in all circuits, self-regulation due to inverse physical connections, thermoelectric energy conversion), which also takes into account the necessity of reserves. The project meets modern nuclear energy safety requirements and is based on the experience of creation and scientific management of atomic energy units and spent fuel elements solutions of nuclear power engineering [6].

3.6. Nuclear subsea gas pumping station

A subsea autonomous nuclear gas pumping station for underwater gas pipelines was laid from the places of offshore natural-gas production to the continent was engineered by JSC CCB ME "RUBIN". The project was developed using the results of research and development performed by corporations of the military-industrial complex of Russia during engineering of main power units of nuclear submarines and surface ships of the Navy. Multiyear experience in the operation of nuclear submarine equipment was additionally used in the course of the project to provide extra safety and ensure the reliability of the facility.

Technical solutions used in the project provide the opportunities for: transportation of the station to the place of installation and operation in the above-water position; immersion of the station at the depth of the main gas pipeline; plugging the station into the gas pipeline through the gas pipeline support and landing platform; long-term and reliable operation (in automatic control mode) as part of the main gas pipeline, creating and maintaining a degree of compression of the pumped gas to ensure the required productivity of the gas pipeline; unplugging of the station from the main gas pipeline and replacing it with a similar one in the event of an failure or after the end of service life; periodic visits to the station by support personnel using an underwater transport vehicle for service of its mechanisms and change of consumables (oil).

The project includes systems and mechanisms that ensure the possibility of a long-term operation of the station in autonomous mode, as well as propulsion systems that provide the station the ability to manoeuvre in an underwater position after bringing it to zero buoyancy. The staff is provided with the possibility to periodically visit the working station for service maintenance. There are living and service rooms for the staff during their stay at the station.

3.7. The atomic icebreaker project 10510 – "Leader"

The 10510 "Leader" nuclear icebreaker is designed for year-round navigation of large-capacity transport vessels with a deadweight of more than 100,000 tons and a width of more than 50 metres all along the Northern Sea Route at an economically effective speed of about 15 knots with an ice thickness of about 2 metres. The unique design of the newest Russian nuclear-powered vessel will support the navigation along the Northern Sea Route even by large-tonnage tankers and dry cargo vessels, which cannot currently use this route. The hulls and the power of 120 MW screws will provide the icebreaker with unprecedented navigation capability in sea ice.

For the "Leader" icebreaker project, Afrikantov OKBM offered a conceptual design for the RITM-400 reactor with a thermal capacity of 315 MW, twice as powerful as the RITM-200 reactor facility installed on the icebreaker project 22220. The required capacity of the nuclear-powered icebreaker "Leader" will be two RITM-400 installations.

4. Conclusion

Based on the overview and analysis, it is possible to draw the following conclusions about the development of nuclear technologies applicable in the Russian Arctic:

1. The key area in the development of nuclear technologies in the Russian Federation is the development of efficient, highly automated reactor facilities for nuclear power plants of small and medium electric power in the range from 6 to 100 MW. Their usage in the Russian Arctic allows providing electric and heat supply to residential and industrial facilities in distant Arctic areas that do not have centralized power supply and have expensive far distance supply of fuel.

2. The early commissioning of the already built floating nuclear power station "Akademik Lomonosov" will significantly increase the energy supply to objects in the town of Pevek in the Chukotka region of Russia. At the same time, an analysis of the overall budget and long-term construction is needed. The design of the station began in 2008, with commissioning planned for 2019. When implementing the subsequent projects of the Floating nuclear power station, all involved organisations should make maximum efforts to shorten the terms of design, construction and commissioning. It is possible to implement the previously developed FNPP project based on the ABV-type reactor facility, as well as the project of a transportable nuclear power plant with the VBER-300 reactor facility based on a stationary ice-resistant platform for power supply of gas production and processing centres in the Arctic.

3. One of the promising areas for the development of nuclear technologies in the Russian Federation is the implementation of high-temperature gas-cooled reactor projects aimed at expanding the energy market in the sphere of heat supply. Produced heat can be used for various industries, including the intensification of oil production, petrochemicals and refining. According to the vision of the developers, it is possible to implement a high efficiency direct gas turbine cycle (~ 50%) with simultaneous use of waste heat for municipal heat supply. Installation of a single reactor unit and helium-cooled gas turbine plant is used to provide a compact self-contained power source for the surface, underwater and ground objects of the Russian Arctic separated from external infrastructure [9].

4. Russia is the undisputed world leader in the construction of civilian nuclear ships. Icebreakers powered by nuclear power plants have been successfully operating in the Arctic Zone of the Russian Federation for a long time, ensuring the safe passage of transport vessels through the Northern sea route. The current practice of the icebreaker fleet involves navigation of vessels on deep water with linear

icebreakers. Icebreakers with limited draft are used in shallow water areas. This requires the transfer of caravans from the linear icebreaker to the icebreaker with a small draft and vice versa, leading to delays and negatively affecting the economic efficiency of cargo transportation. In order to improve the efficiency of the nuclear icebreaker fleet, a project for the construction of a new-generation universal two-vessel nuclear icebreaker (project 22220) is currently being implemented. This will be capable of combining the functions of both linear icebreakers and icebreakers supporting limited-draft shipping. Icebreakers of this type incorporate two of the latest RITM-200 175 MW reactor units each. The inventions, utility models, software and know-how obtained during the development of this reactor system will allow the construction of RITM-400 reactor units with 110-130 MW capacity for the icebreaker (project 10510), which will be able to provide year-round navigation on the NSR.

Intensive research and development in the above areas will preserve and consolidate Russia's position as a leader in the production and use of nuclear reactor facilities. The Russian Arctic has become a testing ground for the application of the most advanced technologies and scientific research, providing a long-term basis for technological development of the whole country and forming scientific and technical grounds for securing of the actual sovereignty of the Russian Federation in the Arctic [10].

5. Acknowledgment

The paper is based on research carried out with the financial support of the grant of the Russian Science Foundation (Project No. 14-38-00009, programme-targeted management of the Russian Arctic Zone development), Peter the Great St. Petersburg Polytechnic University.

References

[1] JSC "Afrikantov" report. Available from: okbm.nnov.ru/english [Accessed 10 May 2018]
[2] JSC "NIKIET" report. Available from: http://www.nikiet.ru [Accessed 10th May 2018]
[3] NRC "Kurchatov Institute" report. Available from: eng.nrcki.ru [Accessed 10th May 2018]
[4] JSC "Atomenergo" report. Available from: http://atomenergo.spb.ru [Accessed 10th May 2018]
[5] Collection of works of laureates of the international competition of scientific, technical and innovative developments aimed at the development of the Arctic continental and self areas *2017 Moscow: Minenergo of Russia, LLC "Technology of Development"* 176
[6] Collection of works of laureates of the international competition of scientific, technical and innovative developments aimed at the development of the Arctic continental and self areas *2016 Moscow: Minenergo of Russia, LLC "Technology of Development"* 128
[7] Collection of works of laureates of the international competition of scientific, technical and innovative developments aimed at the development of the Arctic continental and self areas 2015 *Moscow: Minenergo of Russia, LLC "Technology of Development"* 136
[8] Collection of works of laureates of the international competition of scientific, technical and innovative developments aimed at the development of the Arctic continental and self areas 2014 *Moscow: Minenergo of Russia, LLC "Technology of Development"* 128
[9] Atroshenko S A, Korolyov I A and Didenko N 2016 Evaluation of physico-mechanical properties of high-chromium tool steels modified with Harrington method *Materials Physics and Mechanics* **26 (1)** 26-29
[10] Didenko N I et al 2017 Analysis of convergence-divergence in the development of economic processes in circumpolar countries *SGEM 2017 Political Sciences, Law, Finance, International Relations* 537-545 (Vienna, Austria)

IOP Conf. Series: Earth and Environmental Science **180** (2018) 012021 doi:10.1088/1755-1315/180/1/012021

Human Activity and the Global Temperature of the Planet

D F Skripnuk and E A Samylovskaya

Peter the Great St. Petersburg Polytechnic University, St. Petersburg, Russia

djamilyas@mail.ru

Abstract. This paper reports on the dependence of the average global temperature on those factors that influence. The current assessment of global temperature is also influenced by previous assessments as of this process as well as by various dynamic factors of human activity. Human activity can be evaluated using a set of factors: waste generated by electricity, gas, steam and air conditioning; waste generated by households; global GDP; road freight traffic; fresh surface water extraction; area of forested land. This paper details the important processes in any given natural environment: a) climatological disasters; b) changes in CO2 levels; (c) proportion of greenhouse gases caused by changes to industry; d) proportion of greenhouse gases caused by changes in agriculture; e) changes in global temperatures; f) changes in fresh water reserves g) changes in area of forested land. To theoretically model the effect of human activity on global temperatures, the autoregressive distributed lags (ADL) model was chosen. Current values generated using the ADL model depend both on previous values, and on the current and previous values of other variables. Empirical testing of the model is described and carried out to validate the results.

1. Introduction

The environment comprises both mankind's habitat and the sphere of its activity. Moreover, the environment includes everything that surrounds the human world including both natural and man-made environments. The number of people on the planet, as well as their quality of life and level of environmental conscientiousness, is directly proportionate to changes in the environment. There are more than 7 billion people on earth and the population continues to grow.

Exponential population growth will lead to a greater human impact on the environment. Many existing problems are caused by the fact that energy consumption plays an important role in human life: people cannot exist without consuming energy. Here, energy is understood as any area of human activity that is connected with the production and consumption of energy.

A significant portion of energy consumption can be attributed to energy released during the combustion of organic fossil fuels (oil, coal and gas), leading to a huge amount of greenhouse gases and pollutants being released into the atmosphere [1]. The air temperature of the environment increases and the greenhouse effect develops. The so-called greenhouse effect is caused by the growth in the concentration of greenhouse gases in the atmosphere. The concentration of these gases is increasing as a consequence of the burning of huge amounts of organic fuel (oil, natural gas, coal, firewood, peat, etc.) due to processes in industry, transport, agriculture and residential settlement. Humans are thus responsible for significant increases in the output of greenhouse gases and are constantly destroying natural ecosystems that regulate their concentration. Deforestation is ongoing, and, according to some estimates, 35-40% of forested land worldwide has undergone deforestation in the last millennium alone.

Moreover, almost all of the steppes have been ploughed and natural meadows destroyed. The human influence on global warming is no longer simply a hypothesis, but an established fact.

2. Indicators estimating the state of the environment

The environment is characterised by the set of processes taking place within it. These include the environmental processes analysed in this paper: a) climatological disasters; b) change of CO2 content associated with emissions; c) change in greenhouse gases associated with industrial emissions; d) change in greenhouse gases associated with agricultural emissions; e) global temperature change; f) change in fresh water reserves; g) change in area of forested land.

Climatological Disasters. These are destructive events caused by extreme climates that damage property and pose a threat to human life. The costs of climatological damage may be estimated in terms of millions of US dollars.

CO2 Emissions. Earth's environment is characterised by the existence of CO2 in the atmosphere. The presence of CO2 depends on CO2 emissions. In order to manufacture a product, it is necessary to consume a certain amount of resources (raw materials, electricity, etc.). As a result of such production, carbon dioxide and other harmful substances are released into the atmosphere. The change in CO2 content is estimated using grams of CO2 per unit of production (gr./USD).

Industrial greenhouse gas emissions. The main greenhouse gases are water vapour, carbon dioxide, methane and ozone. Human industrial activity has led to reduction in wooded areas, the main natural absorbers of carbon dioxide. Industrial greenhouse gas emissions are estimated in tons of CO2 equivalent.

Agricultural greenhouse gas emissions. Agriculture directly and indirectly promotes the emission of three main greenhouse gases: carbon dioxide, methane and nitrous oxide. The change in greenhouse gas content is estimated in CO2 nitrogen equivalent tons.

Global Temperature Change. The environment may be characterised in terms of global temperature. The average temperature is calculated annually for each country and the global temperature derived from the arithmetical mean. Rates of temperature increase are determined by the average global temperature values.

fresh surface water extraction. Water is essential for agriculture and food security. One of the main problems preventing sustainable development is currently a lack of water. Fresh water resources are steadily declining. Fresh surface water extraction is quantified in millions of cubic metres.

Deforestation has a detrimental effect on the planet and happens both as a result of human activity and natural processes. The change in forested area may be quantified in terms of km^2.

3. Literature review

The negative impacts of human activity on the environment have been thoroughly studied. Leading scientists have proposed a number of methods that could be used to negate these negative effects. These focus on the need for every person to make the prevention or slowing of ecological decline a priority [2]. In [3], conflict between human beings and the biosphere is investigated.

Other studies focus on problems pertaining to the atmosphere and to the planet including the preservation of water reserves, subsoil, flora and fauna [4-6]. In [7], a correlation between the declining ecological status of the planet and the use and production of electricity is investigated. In [8] the problem of the influence of multinational corporations on the environment is discussed. In [9], the main questions of global climate change, their influences on ecological systems as well as many aspects of the activities of the global community working towards the decrease of anthropogenic impact on the planet's climate are described.

In [10] communication strategies for reducing poverty are studied. One of the most significant sources of greenhouse gases is the livestock sector [11]. Scientists detail empirical estimates of the influence of climate change on the price of electricity [12].

Investment policy decisions are analysed under the influence of climate change. Scientists have developed an innovative approach which combines three physical climatic models (the model of geophysical dynamics, Goddard's model of space research and the model of the climate) with the general equilibrium model of global trade for studying the influence of climate change [13]. The models of human activity influencing the greenhouse effect are also presented [14-17].

4. Analytic methodology for studying factors influencing global temperature change

In this section, the methods used in this study are described and discussed. When analysing the factors influencing the change in global temperature, it is important to take into account the peculiarities of the global temperature change process. The current assessment of global temperature is influenced by: a) the history of the process; b) the various dynamic factors of human activity; c) other environmental processes.

As a theoretical model of the analysis of factors influencing the change in global temperature we chose the autoregressive distributed lags (ADL) model in which the current values of the series depend both on previous values of the series, as well as on the current and previous values of other temporary variables. It should be noted that this model has been chosen for analysis because it considers the influence of "previous" periods on "current" ones. Moreover, the model allows the prediction of an endogenous indicator for "future" periods.

The multiplicative type of ADL model is as follows:

$$Y_t^i = \prod_{k=1}^{m} \left(Y_{t-k}^i\right)^{a_{1k}} \times \prod_{k=0}^{m} \left(X_{1,t-k}^i\right)^{a_{2k}} \times \prod_{k=0}^{m} \left(X_{2,t-k}^i\right)^{a_{3k}} \times \prod_{k=0}^{m} \left(X_{3,t-k}^i\right)^{a_{4k}} \quad (1)$$

Having expressed the left and right part of the equation through logarithms on e basis, we get a linear equation of the form:

$$\ln Y_t^i = a_1 \ln Y_{t-k}^i + a_2 \ln X_{1t}^i + a_3 \ln X_{2t}^i + a_4 \ln X_{3t}^i \quad (2)$$

The multiplicative function is transformed to the additive one.

$$\ln Y_t^i = \sum_{j=1}^{k^1} a_0 \ln Y_{t-j}^i + \sum_{j=1}^{k^2} a_1 \ln X_{1,t-j}^i + \sum_{j=1}^{k^3} a_2 \ln X_{2,t-j}^i + \sum_{j=1}^{k^4} a_3 \ln X_{3,t-j}^i \quad (3)$$

A method for empirically testing the model is developed. The method includes the following main stages. When testing the stationarity of time series variables, the Dickey-Fuller test is used to find the coefficient in the autoregression equation of this form:

$$y_t = a\, y_{t-1} + \varepsilon_t \quad (4)$$

, where y_t — is the time series, ε_t — is the error.

If $|a| < 1$, the series is stationary. If a=1 then the process has a unit root. In this case, the series is non-stationary, it is an integrated time series of the first order.

The testing of exogenous variables' multicollinearity: Testing the exogenous variables' multicollinearity is performed by finding the coefficients of pair correlation between the exogenous variables. When the value of the correlation coefficient is greater than $|0.7|$, one variable of the pair is excluded from further analysis. If $t_{tempirical} > t_{theoretical}$ the received coefficients are significant i.e. selection corresponds to the population.

The choice of endogenous variable lags, which have a strong correlative connection with the variable's value in the last period, as well as testing of the significance of autocorrelation coefficients, is carried out by means of Ljung-Box Q-test. The choice of endogenous variable lag having a strong correlative connection with the same variable value in the last period was carried out by means of autocorrelation coefficients. The significance of autocorrelation coefficients was confirmed using the Lyyunga-Boksa criterion. Based on the results, we chose those lags having a strong correlation with the value of the indicator in the previous period. The choice of exogenetic variables has a strong correlative connection with the value of an endogenous variable in the former period; the significance of correlation coefficients was tested. The composition of the structural form of ADL uses a model of the multiplicative type. The values of the natural logarithms of the endogenous and exogenous variables are derived for the period from 1995 to 2016. New variables were introduced and the structural form of the additive type of ADL model presented.

The determination of model coefficients was carried out using regression analysis. The significance of the regression equation and coefficients of the regression equation were verified. Equations were written out using the multiplicative form. Analysis of the economic signification of exogenetic variable coefficient values was performed.

5. Data and analysis of primary data

The work with the initial data consisted of two stages: a) data collection for selected variables; b) primary processing of endogenous and exogenous variables in accordance with the requirements of the model. The data were collected over a period of time from 1995 to 2016.

The data contain the values of the variables for the corresponding year: endogenous variable; exogenous variable – a background process; exogenous variables are factors of human life whereas exogenous variables are characteristics of other processes. Data were taken from the following sources: Euromonitor Passport Database, http://www.euromonitor.com/; World Bank Open Data, http://data.worldbank.org/; CIA of the USA website ttps://www.cia.gov/.

6. Empirical testing of the model

The check of the time series variables for stationarity using the Dickey-Fuller test showed that all the series are stationary.

All the series in the autoregressive equation, the coefficient $|a| < 1$, the values of the coefficient lie in the range $[0.030 - 0.9791]$.

The checking of exogenetic variables' multicollinearity revealed the existence of a strong connection between some exogenetic variables. Inspection of the correlation coefficients' significance showed that for all factors the condition of calculation $t_{tempirical} > t_{theoretical}$.

$X_5, X_6, X_7, X_8, Y_6, Y_7$ are excluded from the further analysis: waste generated by electricity, gas, steam and air conditioning supply (tonnes); waste generated by households (tonnes); global GDP (US$ per capita); road freight traffic (million net tonne-kilometres); fresh surface water extraction (million m^3); forest land (km^2).

The choice of the endogenous variable lag showed that two prior periods in the last period have a high autocorrelation coefficient of a variable. The check of the significance of autocorrelation coefficients on Q-statistics has confirmed their significance (values of Q-statistics are equal to 16.65940 and 27.51671, it is more than the tabular value $\chi^2_{0.05,2} = 5.991$).

The choice of exogenous variables which have a strong correlative connection with the value of an endogenous variable in the previous period allowed preservation of the exogenous variables for the analysis. Factors of human activity: methane emissions $(k_t\ of\ CO2\ equivalent)$ (X_1) in t and $t-1$ periods; greenhouse gas emissions (tones of CO2 equivalent) (X_3) in t, $t-1$, $t-2$ periods; CO2 emissions from the consumption and flaring of fossil fuels (tonnes) (X_4) in t period; agricultural land (sq. km) (X_9) in t, $t-1$, $t-2$, $t-3$ period and exogenous variables - characteristics of other processes: CO2 Emissions per Unit of Output (grams per USD inconstant prices) (Y_2) in t period; greenhouse gas emissions from Agriculture (tones of CO2 equivalent) (Y_4) in t period. All values of correlation coefficients are significant, $t_{tempirical}$ is in the range $[6.87768\ to\ 8.43441]$ at $t_{theoretical} = 1.753$.

The structural ADL model of a multiplicative type has the form:

$$Y_{5_t} = a_0 \cdot Y_{5_{t-1}}^{a_1} \cdot Y_{5_{t-2}}^{a_2} \cdot x_{1_t}^{b_1} \cdot x_{1_{t-1}}^{b_2} \cdot x_{3_t}^{c_1} \cdot x_{3_{t-1}}^{c_2} \cdot x_{3_{t-2}}^{c_3} \cdot x_{4_t}^{d_1} \cdot x_{9_t}^{e_1} \cdot x_{9_{t-1}}^{e_2} \cdot x_{9_{t-2}}^{e_3} \cdot x_{9_{t-3}}^{e_4} \cdot Y_{2_t}^{f_1} \cdot Y_{4_t}^{g_1} \quad (5)$$

The introduction of new variables allowed the presentation of the structural form ADL models in an additive form:

$$A_{5_t} = a_0 + a_1 A_{5_{t-1}} + a_2 A_{5_{t-2}} + b_1 B_{1_t} + b_2 B_{1_{t-1}} + c_1 B_{3_t} + c_2 B_{3_{t-1}} + c_3 B_{3_{t-2}} + d_1 B_{4_t} + e_1 B_{9_t} + e_2 B_{9_{t-1}} + e_3 B_{9_{t-2}} + e_4 B_{9_{t-3}} + f_1 A_{2_t} + g_1 A_{4_t} \quad (6)$$

The model coefficients are found using the regression analysis:

$$A_{5_t} = -23.6543 - 0.3595 \cdot A_{5_{t-1}} - 0.5790 \cdot A_{5_{t-1}} + 0.5554 \cdot B_{1_t} + 0.5442 B_{1_{t-1}} + 0.5695 B_{3_t} + 0.4643 B_{3_{t-1}} - 0.4387 B_{3_{t-2}} + 0.3057 B_{4_t} + 1.6649 B_{9_t} + 1.6157 B_{9_{t-1}} + 0.9317 B_{9_{t-2}} + 0.9867 B_{9_{t-3}} - 0.3206 A_{2_t} + 0.9846 A_{4_t} \quad (7)$$

The equation in a multiplicative form is written as:

$$Y_{5_t} = -23.6543 \cdot Y_{5_{t-1}}^{-0.3595} \cdot Y_{5_{e-2}}^{-0.5790} \cdot x_{1_t}^{0.5554} \cdot x_{1_{t-1}}^{0.5442} \cdot x_{3_t}^{0.5695} \cdot x_{3_{t-1}}^{0.4643} \cdot x_{3_{t-2}}^{-0.4387} \cdot x_{4_t}^{0.3057} \cdot x_{9_t}^{1.6649} \cdot x_{9_{t-1}}^{1.6157} \cdot x_{9_{t-2}}^{0.9317} \cdot x_{9_{t-3}}^{0.9867} \cdot Y_{2_t}^{-0.3206} \cdot Y_{4_t}^{0.9846} \quad (8)$$

7. Results and Discussion

Following analysis of the model, it is possible to draw some conclusions. The indicator "change in average value of global temperature" in the previous period is weakly influential in the current period. This can be explained by the fact that the population increases every year, involving an increase in the production and consumption of resources. The result of an increase in these indicators is the emission of carbon dioxide into the atmosphere, leading to an increased global warming effect. There is "an effect of accumulation" of CO2 and other greenhouse gases in the atmosphere.

The influence of the "change in average value of global temperature" indicator is increased by the growth of the "methane emissions" indicator. In terms of promoting global warming, methane is the second most prevalent greenhouse gas after CO2. The logic is the same as described above. The annual increase in population leads to growth in the consumption of resources and consequent emissions. The result of the growth of these indicators is the emission of various gases into the atmosphere, leading to a heating effect.

With the growth of the "emissions of greenhouse gases" indicator, the influence of this indicator on "change in average value of global temperature" is increased in the current period. The greater the prevalence of greenhouse gases in the atmosphere, the more severe the resultant global warming effect.

With the growth of the indicator "CO2 emissions caused by consumption and combustion of fossil fuels" the influence on "change in average values of global temperature" is increased. This may indicate that as more resources are consumed, more waste gas is formed and the global warming effect becomes stronger.

The variable "area of land used for agriculture " has a strong influence on the indicator "change in the average value of global temperature". If the indicator "area of land used for agriculture" increases, then first of all considerable areas of woodland (which is the main "processor" of carbon dioxide) are cut down or burned. Furthermore, development of agriculture involves emissions of gases from cattle, etc. This indicator affects the increase in change in global temperature more strongly than the rest.

The variable "CO2 emissions per unit of production" has the opposite effect on "change in average values of global temperature" in the current period. Possibly this is so because of restrictions for emissions introduced by the enterprises which, in turn, promote a change to newer technologies and more rational use of resources.

The influence on "change in average value of global temperature" is increased in the current period due to the growth of the "emissions of greenhouse gases from agriculture" indicator. Agriculture is one of the main causes of climate change. Emissions of greenhouse gases from this kind of activity are significant. These gases concentrate in the atmosphere and thereby promote the global warming effect.

Acknowledgements
This paper is based on research carried out with the financial support of the grant of the Russian Science Foundation (Project No. 14-38-00009). Peter the Great St. Petersburg Polytechnic University.

References
[1] Monk G J and Molnar J M 2011 Energy-Efficient Greenhouses *Horticultural Reviews* **9** 1–52
This reference has two entries but the second one is not numbered (it uses the 'Reference (no number)' style.
[2] Caney S 2012 Just Emissions *Philosophy & Public Affairs* **40 (4)** 255–300
[3] Coetzer K L, Witkowski E and Erasmus B F 2014 Reviewing Biosphere Reserves globally: effective conservation action or bureaucratic label? *Biological Reviews* **89 (1)** 82–104
[4] Didenko N I, Skripnuk D F and Mirolyubova O V 2017 Urbanization and greenhouse gas emissions from Industry *IOP Conf. Series: Earth and Environmental Science* **72** 012014
[5] Atroshenko S A, Korolyov I A and Didenko N 2016 Evaluation of physico-mechanical properties of high-chromium tool steels modified with Harrington method *Materials Physics and Mechanics* **26 (1)** 26–29
[6] Romashkina G F, Didenko N I and Skripnuk D F 2017 Socioeconomic modernization of Russia and its Arctic Regions *Studies on Russian Economic Development* **28 (1)** 22–30
[7] Didenko N and Kunze K 2014 Relationship between Energy Policies and Quality of Life in Russia *WIT Transactions on Ecology and the Environment* **186** p 3–1 (WIT Press)
[8] Sharma S and Ruud A 2003 On the path to sustainability: integrating social dimensions into the research and practice of environmental management *Business Strategy and the Environment* **12 (4)** p 205–214
[9] Dhooge L J 2003 A model and resource guide to incorporating human rights education into the business law classroom *Journal of Legal Studies Education* **21 (1)** p 33–85
[10] Rudenko D Y, Pogodaeva T V and Didenko N I 2015 Poverty alleviation strategies in the Russian arctic zone regions *Mediterranean Journal of Social Sciences* **6 (1)** 32–39
[11] Sakadevan K and Nguyen M-L 2017 Livestock Production and Its Impact on Nutrient Pollution and greenhouse gas emissions *Advances in Agronomy* **141** p 147–184
[12] Véliz K D et al 2017 The effect of climate change on electricity expenditures in Massachusetts *Energy Policy* **106** 1–11
[13] Riener K and Kühn F E 2014 CO2 – Baustein des Lebens und Treiber der globalen Erwärmung *Chemie in unserer Zeit* **48 (4)** 260–268
[14] Ramanathan V, Callis L and Hansen C J 1987 Climate-chemical interactions and effects of changing atmospheric trace gases *Reviews of Geophysics* **25 (7)** 1441–1482
[15] Gunst A 2010 Carbon pollution (greenhouse gas) measurement and reporting *Asia-Pacific Journal of Chemical Engineering* **5 (4)** p 646–656
[16] Speight J G and Singh K 2014 Environmental Aspects *Environmental Management of Energy from Biofuels and Biofeedstocks* (USA) chapter 2 p 53–91
[17] Didenko N I et al 2017 Analysis of convergence-divergence in the development of economic processes in circumpolar countries *SGEM 2017 Political Sciences, Law, Finance, International Relations* 537-545